Sapiens Reinvented:

Saving the Species from a Deadly Evolutionary Flaw

Title: Sapiens Reinvented: Saving the Species from a Deadly Evolutionary Flaw

Author: Jim Loehr
ISBN: 978-1-944927-15-8
Publisher: Kipcart Studio, LLC
Printed in USA
First Printing: 2023

For permissions or further inquiries, please contact:
Jim7loehr@gmail.com

Visit Jim-Loehr.com

DEDICATION

To the countless individuals who have endured unimaginable pain and suffering from senseless wars, genocide and aggression throughout human history.

CONTENTS

INTRODUCTION

Humanity's Dual Nature: A Deep Dive into Our Dark Side

As you turn the pages of this book, my fervent hope is that it will not only capture your attention but also shake the very core of your beliefs. Of the 19 titles I have authored, penning this particular book has been an arduous journey, both emotionally and intellectually. Meticulously researching, pouring over countless documents, and assembling this manuscript has repeatedly stretched my emotional endurance to its limits.

This volume is more than just a collection of words. It's an unvarnished confrontation with the harrowing realities of the darker aspects of human nature. It exposes the cruel ways Homo sapiens have historically treated their kin and serves as an impassioned plea for change to prevent further anguish, torment, and fatalities.

The pioneering work of esteemed scholars like David Livingstone Smith, Richard Wrangham, David Hamburg, Robert Sapolsky, James Waller, Simon Baron-Cohen, among others, laid the foundation for this expose. Their audacity to delve into the painful truths of humanity's cruelty towards its members is commendable. Given the daily barrage of disheartening news and unrelenting stress we face, the natural inclination for many might be to steer clear of such grim topics, like war, genocide, prejudice, ethnic cleansing, and political malfeasance. The pressing question then becomes: Why would anyone willingly immerse themselves in such disconcerting matters?

However, what if understanding our sinister side held the key to transforming it? That is the pivotal question this book urges you to ponder and is the central thrust of our exploration.

Part I offers insights into the intrinsic behavioral tendencies of humanity, tracing our penchant for aggression and violence back to deeply embedded genetic codes. These innate traits have been instrumental in ensuring our species' survival, leading our global population to surpass 8 billion. But the sobering reality is that with nuclear capabilities spread across nine countries, this aggressive predisposition threatens our very existence.

Part II delves into the heart-wrenching ramifications of these genetic predispositions. From the horrors of slavery and the devastation of wars to the chilling realities of genocide and pervasive prejudices, these manifestations of our nature are inherently linked to an instinctive drive: ensuring our own survival and that of our descendants, no matter the moral or ethical cost. Over millennia, these primal urges have molded the moral compass of humanity, influencing how we discern right from wrong.

A particularly alarming manifestation of this predisposition is the propensity to dehumanize those deemed different or threatening. Labeling them as "others" or "outsiders," especially when competing for limited resources, often gives a perceived moral license to commit unspeakable atrocities to safeguard one's tribe, community, or nation.

Part III delves into the complexities of the Homo sapiens brain, attempting to comprehend how such a marvel of evolution can perpetrate heinous acts devoid of remorse, guilt, or self-reproach. The unsettling truth is evident: the incredible adaptability of our brains, while being a source of our evolutionary success, also connects directly to our capacity for cruelty and inhumane actions.

Part IV shines a beacon of hope amidst this bleak analysis. It presents a tangible solution to rectify humanity's inherent flaw, bolstered by rigorous scientific evidence from fields like neuroscience, epigenetics, and social-emotional learning research. The book furnishes a detailed 20-week training regimen premised on the belief that the most promising prospects for positive change lie with today's youth. Thus, it seeks to equip key influencers – parents, educators, and coaches – with tools to nurture a more compassionate, empathetic generation.

Note:

The urgency to act and instigate change on a global scale is palpable. It is my earnest hope that this read will not only inform but also inspire you to join a burgeoning movement, a movement that seeks to supplant humanity's innate violent tendencies with a profound sense of empathy, solidarity, and interconnectedness with every member of our species.

Part I:
The Flaw

The Enigma of Evolution:

The Homo Sapiens' Brain

Around 70,000 years ago, something truly miraculous started happening in the brain of Homo sapiens. An evolutionary upgrade occurred, giving rise to two groundbreaking abilities: the capability of comprehending the inevitability of one's own death and the inevitability of the death of all loved ones. This major leap, made feasible by the expansion of the cortex and neocortex, seemed to be in direct conflict with the genetically coded mandate to survive at all costs and ensure the survival of one's species. Though enlightening, this existential awareness came with its own challenges. It instigated an ongoing search for meaning, purpose, and a deeper understanding of existence. While it is this very quest that led to the development of art, culture, religion, and philosophy, it also sowed seeds of anxiety, aggression, despair, and the fear of the unknown. The shadows of mortality, paradoxically, pushed Homo sapiens both to heights of creation and depths of inhumanity. And this is where the **Flaw** emerges: "In whatever time I have, I must do everything in my power to ensure the survival of my people, my species, my tribe, my nation, etc. Whatever action I must take is, therefore, morally justified."

Journeying back, in 1932, the world was greeted with a profound discovery: an ancient skull from the Skhul Cave in Mount Carmel, Israel. This 90,000-year-old relic belonged to an adult male bearing a striking resemblance to modern humans, characterized by an elevated forehead. This chronicle moves to Europe, where approximately 40,000 years ago, the Cro-Magnon – the predecessors of contemporary humans – made their presence felt. Their existence is attested by intricate cave paintings representing spiritual entities, the beginnings of coherent language, and primitive musical instruments. These artifacts signify the ever-evolving brain of the Homo sapiens and serve as testaments to our insatiable curiosity and hunger for meaning.

The narrative of human evolution is not one of raw power or speed. Indeed, Homo sapiens were neither the strongest nor the swiftest of the species. Yet, they emerged as the apex predators, dwarfing the legacies of Homo erectus and the Neanderthals. The latter species met their downfall, most probably due to a mix of interbreeding and outright extermination by their cognitively superior competitors. This dominance can be attributed to the unparalleled evolution of the Homo sapiens brain, a subject that continues to baffle and intrigue neuroscientists and evolutionists alike.

The brain, a dense web of intricately interwoven neurons, is nothing short of an enigma. How can mere three pounds of this organ yield conscious cognizance, granting the ability for introspection and self-awareness? This phenomenal entity, although constituting a minuscule 2-3% of the total body mass, surprisingly guzzles nearly a quarter of the body's energy when in a resting state.

When one observes the human species in its primal form, it becomes apparent how utterly defenseless they appeared. Stripped of fangs, claws, venom, or any immediate means of defense, Homo sapiens seemed ill-equipped to face the host of perils that Earth had to offer. The external world, rife with predators that outclassed them in strength and speed, posed ominous threats at every turn. Imagine the early Homo sapiens in their primal environment: vast savannas stretched out, teeming with predators. Larger beasts with fearsome claws and fangs dominated these landscapes while venomous creatures slithered through the undergrowth. Homo sapiens, in stark contrast, stood vulnerable —

no natural armor, no deadly fangs, and no venom. Yet, they survived and eventually thrived. The reason? Their unparalleled ability to think, plan, and act. Homo sapiens' trump card lay not in physical prowess but in their unparalleled cognitive abilities.

It was the human brain that was their salvation. This immensely potent neuro-processor afforded them capabilities that no other species could match. It endowed them with skills in abstract thinking, mental reasoning, language proficiency, intricate problem-solving, and foresight. This brain crafted the tools and weapons, evolving them from rudimentary spears to the sophisticated machinery and lethal arsenals of today.

In essence, the journey of Homo sapiens is nothing short of a marvel. In the grand theater of life on Earth, while many species played their parts relying on physical strength, the Homo sapiens chose a different path. They placed their bets on the power of their minds. Today, as we gaze at the stars, create digital wonders, and ponder the mysteries of our existence, it's evident that the gamble of our ancestors paid off. The tapestry of human existence, intricately woven with threads of thought, innovation, and emotion, stands as a testament to the unmatched power of the Homo sapiens' mind.

In the Homo sapiens' journey to becoming Earth's unrivaled rulers, this brain, with its boundless potential, became, tragically, the accelerator of their deadly flaw in the species. From devising ingenious strategies to outwit outsiders and those competing for scarce resources to developing innovative weaponry that could completely vanquish perceived threats, Homo sapiens wielded their cognitive capabilities like an artist brandishing a paintbrush, continuously crafting and reshaping the tapestry of their ability to protect their progeny by any means necessary. Therefore, the tale of Homo sapiens is not just one of triumphs and dominion. It's a tapestry intricately woven with threads of brilliance and frailty, courage and fear, love and loathing. At the very core of this narrative is the realization that the greatest strength of Homo sapiens, their advanced cognitive ability, is also the source of their most profound vulnerabilities.

The same cognitive processes that gave birth to the ability to love and feel empathy also gave birth to fear, anger, aggression, hatred, and revenge. This, combined with the comprehension of resource scarcity, territoriality, and societal hierarchies, led to feelings of jealousy, hostility, and aggression. As humans began to organize into larger groups and civilizations, these negative emotions often played out in wars, conflicts, and power struggles.

As the chapters of the Homo sapiens story continue to unfold, the hope remains that this balance of power and vulnerability, this oscillation between magnificence and inhumanity, will lead not to self-destruction but to an era of enlightenment, empathy, and sustainable progress. Only time will tell if this species, armed with the knowledge of their strengths and weaknesses, can craft a future that embodies the best of their evolutionary legacy.

Evolutionary Psychology:

Coded Priorities in Homo Sapiens' Brains

The Foundations of Human Evolution

Throughout the progression of evolution, mutations in the human brain have been heavily driven by two dominant forces:

1. Self-Preservation (Survival Instinct)
2. Perpetuation of one's Progeny

Nature prioritizes survival. Thus, behavioral or cognitive adaptations that enhanced the chances of an individual surviving to reproductive age and ensuring the survival of their offspring became ingrained in our DNA. This evolutionary paradigm set the cornerstone for human moral reasoning. The challenge of survival and the drive to protect the next generation introduced man to the notions of right and wrong, painting the backdrop of the complex moral landscape we navigate today.

The Dawn of Social Humanoids

During the nascent stages of human evolution, survival was often a communal effort. The predecessors of Homo sapiens recognized the value of strength in numbers. By banding together in tribes and clans, early humans could protect against predators, share resources, and raise the young in a safer environment.

However, this clan-based survival strategy had its own challenges. The world was vast, and every unfamiliar face posed a potential threat. Those who didn't look, act, or speak the same, or those who worshiped different gods, became natural outsiders. Their intentions were unknown, and in a world where every moment was a game of survival, the default reaction to unfamiliarity was suspicion, if not hostility.

This communal focus on survival also meant that empathy and compassion were predominantly reserved for the 'in-group'. External groups were often dehumanized to justify acts of violence against them. In essence, if an action, no matter how brutal, assured the survival of the tribe, it was often deemed necessary.

Sapiens and the Moral Dilemma

This evolutionary background gave rise to what many might consider Homo sapiens' greatest moral flaw: an unwavering drive for self-preservation at almost any cost. The lens of evolution makes this flaw understandable, but in our modern world, it presents grave ethical challenges.

The drive to ensure one's survival and the survival of one's lineage became so embedded in human nature that it manifested in acts that are considered morally reprehensible today. As societies grew more complex, those in power harnessed and manipulated this instinct to justify acts from theft to even genocide.

David Hamburg, in his seminal work “Preventing Genocide,” touched upon this innate behavior, highlighting that when social norms erode, humans often revert to primal survival tactics. This reversion can be so extreme that entire groups can be dehumanized, with catastrophic consequences.

Depiction of a prehistoric community gathering. Image generated by Midjourney (2023).

Brain Science and Aggression

The human brain, as intricate as it is, houses areas that control our primal instincts. Konrad Lorenz, famed for his work in ethology, contended that aggression is one such instinct honed over millennia to enhance survival. Walter Hess furthered this understanding by identifying the "Lizard Brain" or the diencephalon, particularly the hypothalamus, as a hub for aggressive behaviors.

Aggression itself is multifaceted. As identified by Richard Wrangham, proactive aggression, calculated and premeditated, arises from different neural pathways than reactive aggression, which is often spontaneous and driven by heightened emotions. Yet, both types are anchored in the same evolutionary regions of the brain.

Here are examples of both:

Proactive Aggression	**Reactive Aggression**
All offensive wars	Aggression against an attacker
Genocide	Road rage
Serial killings	Spouse abuse
Death camps	Aggression in defense of your home or possessions.
Ambushes	Kicking your dog because it bit you.
Sniper kills	Aggression in defense of an attack on your friend.
Firing squads	Spontaneous attack on your boss immediately after being fired.

Religious prejudice	Violent attack on someone after a derogatory comment about your girlfriend.
Sexual Predators	Throwing a piece of china against the wall because you just got bad news.
Dehumanization in any form	Destroying a tennis racquet or golf club in response to a costly mistake.
Slavery	Getting into a vicious fight over losing a soccer game.
Revenge killings	Attacking your neighbor for playing loud music that kept you awake all night.

The Chimpanzee-Homo Sapiens Dichotomy

Our primate relatives, the chimpanzees, epitomize the essence of reactive aggression. When confronted, threatened, or put in a situation where survival is at stake, they react. This immediate, visceral response is primarily rooted in the 'fight or flight' mechanism. However, Homo sapiens, with their cognitive advancement, ushered in a nuanced form of aggression: proactive aggression.

Whereas reactive aggression is instinctive and immediate, proactive aggression is calculated, deliberate, and often premeditated. This transition from being purely reactive to becoming capable of preplanned aggression catapulted Homo sapiens from being a defensive species to a predatory one. This shift in behavior wasn't just about enhanced aggression; it reflected an evolutionary advantage provided by improved language capabilities, abstract reasoning, and sophisticated strategic planning.

A threatened chimpanzee expressing reactive aggression. Image generated by Midjourney (2023).

Historical Evidence of Proactive Aggression

The annals of history, both ancient and recent, are peppered with evidence of this form of aggression. The discovery of mass graves at Schoneck-Kilianstadten near Frankfurt and the Tollense River site in northern Germany attest to this. These sites, dating back thousands of years, are silent witnesses to violent confrontations, perhaps driven by territorial disputes, resource scarcity, or mere domination.

One can infer from such findings that these were not spontaneous clashes; they bore signs of strategic planning, implying a certain level of abstraction and foresight. This degree of premeditation—whether in territorial annexations, mass murders, or wars—was made feasible by the cognitive leaps manifested in the development of sapiens' cortex and neocortex.

Understanding Gender Disparities in Aggression

Elizabeth Cashdan's "Evolutionary Perspectives on Human Aggression" sheds light on the evolutionary trajectory of human violence. It posits that our aggressive tendencies have been shaped to be more adaptable to external threats. Interestingly, this adaptability has exhibited a marked gender disparity. Historically and contemporarily, males have been the predominant perpetrators of violence, with their involvement in violent acts outnumbering females by a significant margin.

This disparity raises intriguing questions. Why is male aggression so pronounced? While several neurobiological hypotheses have been advanced, ranging from hormonal differences to brain structure variations, a definitive explanation remains elusive. Nevertheless, this phenomenon cannot be denied. The world has witnessed male-led violence on an unparalleled scale, with terrorism being a contemporary exemplar.

The Pantheon of Aggressive Leaders

Throughout history, various figures stand out for their role in the mass perpetration of violence and cruelty. From Julius Caesar's conquests to Pol Pot's genocidal regime, the thread that binds these leaders is not just their shared culpability in widespread death and suffering but also their gender.

Each of these men, whether driven by ideology, thirst for power, or even perceived necessity, has caused the demise of millions. They personify the lethal blend of Homo sapiens' intrinsic aggression and their evolved capacity for strategy and abstract thought. This combination, under certain circumstances, can birth a super-human predator.

The Paradox of Human Nature

Richard Wrangham's "The Goodness Paradox" encapsulates the duality of human nature. Homo sapiens, with their unparalleled intelligence, harbor within them a complex spectrum of moral tendencies. At one end of this spectrum lies the capability for unprecedented cruelty, while at the other, the potential for profound kindness.

It's crucial to ponder: What drives this dichotomy? Is it purely evolutionary, where aggression was essential for survival and dominance, or is it a by-product of societal constructs, ideologies, and power dynamics?

As Homo sapiens, we're heirs to a legacy that oscillates between extraordinary benevolence and chilling malevolence. It is crucial to understand the evolutionary underpinnings of our aggressive tendencies and recognize the factors that amplify them.

While proactive aggression might have been a survival-enhancing strategy in prehistoric times, its unchecked manifestation in our modern, interconnected world can lead to devastation. As we grapple with this innate duality, it's imperative to cultivate awareness, empathy, and restraint, ensuring that the advanced faculties of strategic planning and abstract thinking are used constructively, steering us towards a more harmonious coexistence.

Conclusion: Understanding Our Past to Shape Our Future

Homo sapiens have traversed a long evolutionary journey. The behavioral codes ingrained in our DNA have been pivotal in our survival and dominance on Earth. However, as we've evolved culturally and socially, it's become imperative to understand and challenge some of these primal instincts, especially when they manifest in ways that can harm our global community.

While our past has undoubtedly shaped us, recognizing, understanding, and addressing the moral implications of our evolutionary history is crucial. As the world grows more interconnected, the lines dividing 'us' from 'them' blur, and we become responsible for ensuring that the old survival codes don't jeopardize our collective future.

In 1963, The Bronx Zoo had an Exhibit called:

“The Most Dangerous Animal in the World”

It was a Mirror!

Part II:
The Consequences

The Moral Failing of Homo Sapiens:

A Root Cause of History's Greatest Tragedies and Human Suffering

Prejudice: Evolutionary Beginnings

Throughout the evolutionary process, prejudice has played a sinister yet integral role in the survival strategies of our ancestors. When resources were scarce, distinguishing between "us" and "them" was not just a sociocultural practice but a pivotal survival mechanism. Minute differences in physical appearance, like skin color, hair type, or even the nuances in the manner of speaking, marked the difference between potential allies and potential threats.

From a survival perspective, it made logical sense. In a world where food, shelter, and security were never guaranteed, these stark contrasts between different groups of people acted as a precautionary measure. Any potential threat to the security of the community had to be identified and dealt with. This instinctual behavior favored groups that were more homogeneous and mistrustful of outsiders, as they were less likely to face betrayal or surprise attacks from within their own ranks. Over time, these instincts solidified into deeply ingrained societal behaviors.

The Mechanisms of Prejudice

Prejudice manifests itself in many forms, but its core remains consistent: it is a hostile attitude directed towards others based on perceived differences, often seen as negative or objectionable. These differences, magnified by fear and ignorance, become justifications for mistreatment. The use of dehumanizing language further serves this narrative. By referring to others as "savages," "vermin," or any other derogatory term, it becomes psychologically easier to mistreat or even eliminate them. It provides a cushion against empathy, making it simpler to disengage from the moral consequences of one's actions.

In this light, prejudice becomes a tool, a mechanism that allows one group to maintain dominance over another. When one considers the moral code of Homo sapiens, which has historically sanctioned actions that favor survival and prosperity, the lack of remorse or guilt in the face of prejudice-fueled actions becomes more comprehensible.

Nature or Nurture? The Genesis of Prejudice

The debate about the origin of prejudice begs the question: is this an inherent trait or a learned behavior? While no concrete evidence suggests that prejudice is hardwired in our genes, evolutionary history showcases a clear trajectory towards favoring one's in-group, especially when survival is at stake. In that sense, the evolutionary mandate to survive could be seen as indirectly encouraging prejudice as it champions the purity and security of one's own group.

However, it's important to understand that this instinctual need to belong to a group and be wary of outsiders doesn't directly translate to the more malicious prejudices we witness in society today. In their current forms, most prejudices are not innate responses but rather learned behaviors. Cultural norms, societal structures, historical events, and individual experiences play massive roles in shaping an individual's prejudices.

From a young age, individuals are bombarded with subtle (and not-so-subtle) messages from family, peers, media, and institutions, which can either foster or challenge prejudicial beliefs. Consequently, prejudices evolve over time, influenced by changing societal norms, political climates, and personal experiences.

The Societal Consequences of Prejudice

In societal constructs, prejudice leads to divisions. These divisions can be seen in social hierarchies, discriminatory laws, economic disparities, and access to basic human rights. Often, prejudices can become so deeply embedded in a culture that they are hardly recognized as prejudices anymore but are accepted as 'the way things are'. This acceptance further perpetuates a cycle of discrimination and bias.

Prejudices, when left unchecked, can lead to severe societal consequences, such as racism, sexism, and other forms of systemic discrimination. Entire communities can be marginalized, leading to cycles of poverty, lack of education, and limited access to resources. In extreme cases, these prejudices can escalate to violent persecution, ethnic cleansing, and even genocide.

Combating Prejudice

Given the pervasive nature of prejudice and its potential consequences, combatting it is paramount. To do so requires both individual and collective efforts. It starts with introspection and acknowledgment. Recognizing one's own biases and actively challenging them is the first step.

Educational institutions play a pivotal role in molding young minds. Curricula that promote diversity, inclusivity, and a genuine understanding of different cultures can go a long way in dismantling inherited prejudices.

In conclusion, while the roots of prejudice trace back to the evolutionary survival instincts of early man, it has evolved into a complex societal issue with dark and ominous consequences. Recognizing, challenging, and educating against prejudice is essential for building a more inclusive, empathetic, and harmonious world.

Slavery: A Deep-Rooted Stain on Humanity's History

Slavery, a word that evokes strong emotions, is a reprehensible and abominable practice that has marred human civilization for thousands of years. Rooted in a desire for cheap labor and economic advantage, it has thrived on the degradation and dehumanization of millions. The very foundation of slavery relies on viewing human beings as nothing more than assets or livestock, which can be bought, sold, used, and discarded as per the owner's whims.

The magnitude of the brutality and scale of slavery, especially in the Americas, is staggering. Estimates suggest that, within the confines of the United States alone, over 10 million souls endured the shackles of slavery, contributing an unimaginable 410 billion hours of forced, uncompensated labor. Their blood, sweat, and tears bolstered the prosperity of a nation, even as their basic human rights and dignity were stripped away.

However, the roots of this dark period in American history can be traced back across the Atlantic. Between 1525 and 1866, the Trans-Atlantic slave trade, one of the most brutal and extensive slave-trading routes in history, saw an estimated 12 million Africans being forcibly removed from their homelands. These individuals were packed onto ships under inhumane conditions, with each ship becoming a floating chamber of horrors. The fact that over 26% of this human cargo consisted of children paints an even grimmer picture of the ordeal. Tragically, more than a tenth of these individuals, overwhelmed by the harsh conditions, disease, or sheer despair, never made it to the New World, their lives snuffed out in the cold, dark depths of the ocean.

The rationale behind this unfathomable cruelty lies in the very process of dehumanization that slavery thrives on. Especially in the context of African slavery, there existed a deeply entrenched belief, even among the educated class, that black individuals lacked souls. By perpetuating this myth, black slaves were relegated to the same status as any other beast of burden, thus justifying any form of cruelty meted out to them. This not only allowed for their brutal treatment but also normalized it.

"Slave Auctions in Richmond, Virginia," George Henry Andrews, The Illustrated London News, vol. 38 (Feb. 16, 1861), p.139. Public domain.

The historical tapestry of slavery stretches far and wide, with its tendrils reaching almost every corner of the globe. As far back as the 18th century BC, we find traces of this heinous practice. From the ancient city of Babylon to the iconic Greek states of Sparta and Athens, slavery was an entrenched institution. Throughout the Mediterranean, during the turbulent times of the Middle Ages, countries like Egypt, Portugal, and Spain witnessed the rise and fall of slave empires. The West Indies, too, was built on the back of slave labor. Surprisingly, even in the Americas, long before the European settlers arrived, indigenous tribes like the American Indians were not strangers to the concept, as they also engaged in buying and selling slaves.

In light of this vast history, one of the most unsettling revelations comes from David Livingstone Smith, the author of "On Inhumanity." Smith posits that the majority of individuals who directly or indirectly benefited from the institution of slavery were ordinary people. These were not monsters or sociopaths but regular citizens with typical moral compasses. This begs the question: how could individuals with normal moral sensibilities participate in or benefit from such a heinous practice?

The uncomfortable answer lies in the power of labels. Once a group is labeled as 'subhuman', it becomes remarkably easy for aggression and exploitation to occur. This dehumanization provides a psychological buffer, allowing individuals to reconcile their actions with their morals. The 'othering' of slaves made it possible for ordinary people to turn a blind eye to the atrocities committed, as they were happening to those deemed less than human.

In reflection, slavery stands as a testament to the depths of cruelty that humans can descend to when motivated by greed, power, and prejudice. The sheer scale and brutality of the practice serve as a haunting reminder of the importance of upholding the principles of equality, justice, and humanity. It's imperative for current and future generations to remember this dark chapter in history, to ensure that such a tragedy is never repeated, and to work towards a world where every individual's inherent worth and dignity are recognized and cherished.

WAR: The Perpetual Scourge of Homo Sapiens

The annals of human history are awash with the colors of war. Far too often, these colors are red, symbolizing the bloodshed that punctuates our collective past. Wars have persistently been fought over territory, resources, religious differences, economic benefit, revenge, and, most poignantly, for self-defense. The irony is palpable. While the very crux of initiating a war may be the inherent desire to protect one's kin and kind, the repercussions of this instinct lead to the obliteration of countless other lives.

Wars necessitate a certain detachment from the enemy's humanity. This detachment, or dehumanization, renders the enemy as lesser beings. Consequently, the unspeakable horrors of war – the violence, brutality, and massacres – are meted out with a twisted moral justification. This cycle has repeated time and time again, evident in both ancient and modern conflicts.

Battle of Gettysburg Casualties, U.S. Civil War (1863). United States Library of Congress's Prints and Photographs division. Public domain.

Perhaps the most tragic aspect of war is the magnitude of the toll it exacts, not just in terms of life but also the trauma that reverberates through generations. While the Homo sapiens brain has evolved to remarkable extents, enabling profound thought, creativity, and innovation, it remains astoundingly inadequate when tasked with grasping the sheer devastation of warfare. Beyond the obvious battle-induced casualties, there's an overwhelming ripple effect: civilians caught in the crossfire, those lost and never found, the countless who succumb to famine, diseases that arise post-war, and a haunting increase in deaths caused by post-war crime, depression, aggression, and vengeance.

The history of our species is intricately laced with wars. Stretching back thousands of years, we find that our ancestors, much like us, were embroiled in relentless conflicts. From ancient wars fought with rudimentary weapons to contemporary wars involving sophisticated warfare technologies, the intent and outcome have remained eerily consistent: territorial conquest, control over resources, and overwhelming death tolls.

To grasp the extent of our belligerent tendencies, consider the following grim catalog of wars throughout history, each accounting for deaths over a million:

1. Chinese Warring States (475 B.C.)
   More than 1,500,000 deaths

2. Punic Wars (264 B.C.)
   Between 1,200,000 and 1,800,000 deaths

3. Roman Civil Wars (91 B.C.)
   More than 3,000,000 deaths

4. Three Kingdoms War (184-279)
   More than 35,000,000 deaths

5. Arab-Byzantine Wars (629-1050)
   More than 2,00,000

6. An Lushan Rebellion (755-762)
   More than 13,000,000 deaths

7. Crusades (1095-1290)
   More than 1,000,000 deaths

8. Mongol Invasions and Conquests (1206-1369)
   More than 30,000,000 deaths

9. Hundred Years War (1337-1450)
   More than 2,300,000 deaths

10. The Conquest of Timur (1370-1405)
    More than 8,000,000 deaths

11. Spanish Conquest of the Aztec Empire (1519-1603)
    More than 24,000,000 deaths

12. Spanish Conquest of Yucatan (1519-1595)
    More than 1,300,000 deaths

13. French Wars of Religion (1562-1598)
    More than 2,000,000 deaths

14. Thirty Years War (1618-1648)
    More than 4,000,000 deaths

15. Mughal-Maratha Wars (1658-1707)
    More than 5,000,000 deaths

16. Napoleonic Wars (1803-1816)
    More than 3,500,000 deaths

17. Taiping Rebellion (1850-1864)
    More than 20,000,000 deaths

18. Miao Rebellion (1854-1873)
    More than 4,500,000 deaths

19. Dungan Revolt (1862-1878)
    More than 8,000,000 deaths

20. World War I (1914-1918)
    More than 16,000,000 deaths

21. Russian Civil War (1917-1922)
    More than 5,000,000 deaths

22. Chinese Civil War (1927-1948)
    More than 8,000,000 deaths

23. Second Sino-Japanese War (1937-1945)
    More than 20,000,000 deaths

24. World War II (1939-1945)
    More than 80,000,000 deaths

25. Soviet-Japanese War (1945)
    More than 32,000,000 deaths (numbers included in World War II estimates)

26. Korean War (1950-1953)
    More than 1,500,000 deaths

27. Vietnam War (1955-1975)
    More than 1,300,000 deaths

28. Afghanistan Conflict (1978- present)
    More than 1,400,000 deaths

29. Second Congo War (1998-2003)
    More than 2,500,000 deaths

Fast forward to the present, and it's heart-wrenching to note that the Russo-Ukrainian War is actively inscribing its own chapter in this already blood-soaked chronicle. The statistics emerging from this conflict – hundreds of thousands dead, millions displaced, and a staggering economic cost – are not merely numbers but represent human lives and livelihoods. Additionally, on October 7, 2023, Hamas terrorists crossed into Israel by land, sea, and air, brutally slaughtering over 1,000 men, women, and children attending a music festival in the Gaza Strip.

It's a harrowing contemplation, the fact that our species, capable of great love, innovation, and progression, is equally capable of unfathomable cruelty. This juxtaposition of our capacities for creation and destruction underscores a haunting paradox. While we celebrate our advancements in art, science, and technology, we simultaneously manifest as the most formidable predators, unparalleled by any other species in our propensity for self-inflicted annihilation.

The question that looms large is: Are we truly the pinnacle of evolution, or are we, in essence, still primitive animals driven by primal urges, merely cloaked in sophisticated veneers? Our intellect is undisputed, but our moral evolution seems to lag considerably behind. The canvas of history paints a conflicted portrait of a brilliant and barbaric species.

At the heart of this dichotomy is our biological drive to survive and reproduce – the root cause of most conflicts. This innate impulse can be seen as a double-edged sword. On one hand, it propels growth, innovation, and progress; on the other, it fuels aggression, jealousy, and war. From an evolutionary standpoint, Homo sapiens might view their actions as both morally acceptable and biologically mandated. Yet, if unchecked, this imperative threatens to undermine the survival it seeks to ensure.

In summation, the narrative of war is intertwined with the narrative of our species. As we stride forward into the future, the true test for Homo sapiens will be our ability to evolve beyond our baser instincts, learn from our history, and genuinely prioritize peace over conflict. Only then can we truly claim to be the most evolved species on Earth.

Genocide: Understanding the Rwandan Crisis and Beyond

History has borne witness to numerous instances where entire communities have been wiped out. However, the Rwandan Genocide stands out as one of the most poignant examples of human-induced tragedy. The events that unfolded in Rwanda during the '90s are a grim reminder of the depths to which human nature can sink, influenced by tribalism, political maneuvering, and unchecked prejudice.

To comprehend the gravity of the situation, one must first examine the intricacies of Rwandan history. The Hutus and Tutsis, though largely indistinguishable in terms of language or culture, were historically differentiated based on socioeconomic class. However, the colonial administration, particularly during the Belgian rule, institutionalized these differences by issuing ethnic identification cards, creating a deep-seated rift between the two groups. This division was not merely social or cultural but became deeply ingrained in the nation's political psyche.

The "Bahutu Manifesto" of 1957 catalyzed this existing division. Advocating for a political shift in power based on population percentages or "statistical law" entrenched the idea that the majority Hutus should naturally have political dominance. Over the years, this rhetoric escalated and evolved, leading to an increasingly volatile political landscape.

Then, in April 1994, an event occurred that would set off a powder keg of tensions: the assassination of the Burundian President, Cyprien Ntaryamira. What followed was an unrelenting onslaught against the Tutsi community. The scale and speed of the violence were staggering. As death tolls rose, the nation plunged into a state of anarchic frenzy. Fueled by propaganda, military directives, and societal pressures, ordinary individuals became participants in one of history's darkest chapters.

An aspect that's particularly distressing is the labeling of the Tutsis as "inyenzi" or cockroaches. Such dehumanization is a tried and tested tool in the arsenal of genocidal regimes. By reducing the targeted group to vermin or pests, the moral boundary preventing mass murder gets eroded. This labeling psychologically enables the perpetrators to view their actions as a mere extermination process rather than the mass killing of fellow human beings.

The formation of youth militias further underlines the manipulation and indoctrination of the younger generation. Impressionable minds were molded to see their Tutsi neighbors, classmates, and sometimes even family members as the enemy. The unchecked spread of hateful propaganda exacerbated the situation, with local media outlets like Radio Télévision Libre des Mille Collines playing a central role in fanning the flames of hatred.

The question remains: What drives ordinary people to commit such heinous acts? This isn't just a question for Rwanda but for the broader study of human behavior. As James Walker's conclusion suggests, the potential for such darkness exists in ordinary individuals. A combination of factors, including fear, propaganda, societal pressure, and a history of prejudice, can push individuals to commit unthinkable acts. When survival instincts are manipulated and weaponized, humans can justify unimaginable atrocities.

Moreover, internalized and perpetuated stereotypes can play a perilous role in influencing collective behavior. David Hamburg's assertion resonates with this sentiment. Genocides are not spontaneous; they are often the result of long-standing prejudices combined with political manipulation and societal circumstances.

The Rwandan Genocide serves as a testament to the horrors humanity can inflict upon itself. It is a stark reminder of the importance of education, awareness, and open dialogue. Societies must actively work to eradicate prejudices, promote understanding, and challenge divisive narratives. Only through a collective effort can we hope to prevent future genocides and ensure that past tragedies are not repeated.

Picture of skulls recovered from a massacre site in Rwanda, Africa. Taken during the official visit of US Rep. Frank Wolf, 2001 • Public domain.

Hitler's Final Solution: A Distorted Morality

The Holocaust, a word synonymous with unimaginable horror and cruelty, stands out in human history as one of its darkest epochs. At the center of this grim tale was Adolf Hitler, who hinged his vile ideologies on a warped interpretation of genetic purity and racial supremacy.

To truly understand the driving force behind Hitler's 'Final Solution', we need to delve deep into the psyche of Nazi Germany and its Führer. Hitler propagated a moral justification for his intent to purge Europe of its Jewish population. He held that Jews posed a biological threat to humanity and that by eradicating them, he was addressing an evolutionary need to preserve the purity of the human race. In this perspective, he imagined himself as a savior of sorts, shielding the world from what he perceived as a 'genetic aberration'.

However, Hitler's radical beliefs did not emerge from a vacuum. They were deeply rooted in long-standing anti-Semitic sentiments prevalent in Europe, dating back centuries. These sentiments were further aggravated by the aftermath of World War I, the Treaty of Versailles, and the financial downturn during the Weimar Republic, making Jews convenient scapegoats for Germany's woes.

Chełmno extermination camp established during World War II by Nazi Germany on the territory of occupied Poland for the purpose of killing Jews. Photographer unknown, 1942. WIkipedia. Public domain.

For Hitler's plan to be successful, he needed widespread collaboration and complicity. Propaganda played a significant role in ensuring this. The Jew was consistently painted as a parasite, as an enemy, and even as a vermin to be exterminated. By continuously dehumanizing them, Hitler and the Nazi propaganda machine created an environment where millions could be subjected to unspeakable horrors, all the while being convinced that it was for the greater good.

Claudia Koonz, in her insightful book *The Nazi Conscience*, points out the paradox that led to the horrors of Auschwitz and other extermination camps. The atrocities weren't driven by an inherent evilness but by a twisted sense of righteousness. In the Nazi narrative, they were protecting the future of the Germanic race and, by extension, humanity.

There's an evolutionary argument to be made here. Throughout history, humans have shown a propensity to protect and promote their own kind, sometimes at the expense of others. This might stem from an instinctual drive for survival and propagation. However, when these base instincts intertwine with skewed ideologies, the results can be catastrophic.

The stark dichotomy of the German soldiers' behavior during this era is a haunting testament to this skewed worldview. Many of these soldiers, who during the day were participants in ghastly acts, could seamlessly transition into their roles as fathers and husbands by night. This dissonance points to a profound cognitive disjunction, allowing them to compartmentalize the horrors they were a part of and their everyday lives. Hitler's regime had, through relentless propaganda, effectively subverted their moral compasses.

The Eichmann trial, one of the post-war efforts to bring Nazi war criminals to justice, shed light on this unsettling reality. Adolf

Eichmann, often called the 'architect of the Holocaust', was found to be psychologically 'normal' after a psychiatric evaluation. Thomas Merton's reflection on this fact underscores the disturbing realization that the capacity for immense evil can exist in seemingly ordinary individuals.

The Holocaust and Hitler's 'Final Solution' force humanity to confront uncomfortable truths about the depths we can descend when led astray by distorted ideologies and beliefs. The haunting images of concentration camps, gas chambers, and mass graves are a grim testament to what happens when prejudice is left unchecked, when propaganda replaces truth, and when the worst of our instincts is leveraged for nefarious ends.

To remember and understand the Holocaust isn't just about paying respect to its millions of victims; it's about ensuring that we remain vigilant, ensuring that we educate future generations about the dangers of unchecked prejudice, and ensuring that such a tragedy is never repeated. The lessons from this dark chapter in human history must guide our collective conscience, ensuring that we prioritize empathy, understanding, and truth over divisive and prejudiced narratives.

Religious Wars and Human Evolution

The tapestry of human history is intricately woven with threads of belief, faith, and often, conflict. The origins of religious thought are deeply entrenched in the evolutionary journey of Homo sapiens. Archaeological discoveries of ancient burial grounds suggest that religious ideas possibly emerged during the Middle and Lower Paleolithic periods, hinting at the existence of a spiritual understanding some 300,000 years ago.

By the Middle Ages, which spanned between the 5th and 15th centuries, modern-day religions such as Christianity, Buddhism, and Islam had carved out their domains. This epoch also witnessed significant religious clashes, like the Byzantine-Arab wars, the Crusades, and the Islamic conquest of Persia. But to comprehend why religious wars persist, we must first dissect the human psyche's relationship with mortality and the role religion plays in resolving this existential enigma.

The cognitive prowess of Homo sapiens sets our species apart. While animals operate predominantly based on instinct and immediate survival needs, humans possess the profound (and often burdensome) capacity to contemplate their mortality. No other creature on Earth loses sleep over the inescapable truth of death. Yet, the thought haunts the human mind: we know that every heartbeat brings us closer to our inevitable end.

Smoke rises from the site of the World Trade Center Tuesday, Sept. 11, 2001. U.S. National Archives' Local Identifier: P7127-23. Public domain.

Such awareness breeds what might best be referred to as a *spiritual crisis.* Our biological imperative drives us to procreate and ensure the continuity of our lineage, yet we're also plagued by the knowledge that death awaits, rendering our individual existences seemingly futile. Religion, with its transcendent narratives, offers solace. From the mummification rituals of ancient Egypt to the Christian idea of heaven and the Hindu belief in reincarnation, religions propose solutions to the death conundrum. These spiritual narratives offer hope that life extends beyond the physical realm, providing meaning to human existence, pain, and suffering.

However, religion's attempt to answer life's most daunting questions has inadvertently sown seeds of discord. While religions stem from beliefs, they often metamorphose into irrefutable truths for their followers. Herein lies the danger: when personal beliefs are seen as divine decrees, unwavering adherence becomes the norm, and those who dissent are often viewed as heretics or infidels.

Monotheistic religions champion the existence of a singular deity, while polytheistic faiths celebrate a pantheon of gods. Yet, the existence or nature of such divine entities remains elusive to empirical evidence, firmly anchoring these ideas in the domain of belief. But the moment beliefs transition into incontrovertible truths, the ground is set for potential conflict. For followers, these aren't mere perspectives but divine edicts. Thus, non-believers or those following different beliefs become adversaries, often deemed inferior or misguided.

This religious prejudice stems from a sense of divine exclusivity. If believers perceive themselves as the chosen ones, it devalues and dehumanizes non-believers. This divisive mindset underpins many religious wars and conflicts throughout history. It's a manifestation of Homo sapiens' moral flaw: the propensity to value one's in-group while ostracizing the "other."

Despite its pitfalls, one cannot deny the profound contributions of religion to human civilization. It has inspired art, shaped cultures, and even propelled scientific inquiry. Religious teachings have imbued societies with moral frameworks, and religious institutions have served as bastions of learning, charity, and community-building.

However, the intent here isn't to sanctify or vilify religion but to introspect upon its dual nature. While it offers existential solace and societal structure, it can also be a catalyst for division and conflict.

In examining the phenomenon of religious wars, one traverses the evolution of Homo sapiens from primordial beings grappling with the understanding of mortality to civilizations clashing in the name of the divine. The duality of religion — as both a unifying force and a divisive tool — mirrors the very nature of humanity itself: capable of immense love and understanding but also profound prejudice and violence. As our understanding of ourselves and the universe expands, embracing beliefs that unite rather than divide becomes imperative, ensuring that the narrative of human history is one of progress, inclusivity, and peace.

Inner City Gangs

Gangs have evolved as intricate social structures within modern society, with an influence stretching across the world. In the United States, this phenomenon is starkly evident. Statistics from the National Drug Intelligence Center reveal an astonishing figure of 20,000 gangs and over a million members nationwide. These numbers aren't mere data points but represent individuals whose life trajectories have intersected with the gang life. In cities like Chicago, the ubiquity of gangs, with an estimated 70-75 gangs harboring over 100,000 members, highlights a pressing urban crisis.

To understand the rise and pervasiveness of gangs, it is crucial to delve into the nuances of their characteristics, their members, and the societal factors that contribute to their formation. Gangs differ from other social groups in their propensity for extreme violence. This violence serves as a survival tool and a symbol of identity. While gang members can range in age, the average age is pegged at 17, situating gangs as a predominantly youth phenomenon. This youth focus is concerning, as it underscores the dire need for intervention strategies targeted at younger populations.

Scholars and social researchers often link the allure of gangs to systemic societal issues. The precarious balance of socioeconomic conditions in many urban centers—marked by high unemployment, deeply entrenched institutional racism, and abject poverty—creates fertile ground for gangs to take root. This milieu becomes a potent cocktail, driving the youth, especially those from marginalized communities, into the waiting arms of gangs.

The coded mandate of survival underscores these choices. From an evolutionary standpoint, humans have always been driven by the need to survive, and gangs, in the landscape of urban decay, appear as a viable means to that end. The gang becomes a way out—a beacon of hope in a bleak world, promising the alluring prospects of money, respect, and above all, survival.

The underground economy that gangs operate within offers tangible benefits. For a young individual trapped in the endless cycle of poverty, gangs can provide the means to obtain essential necessities like food, shelter, and security. Beyond just economics, gangs offer an identity—a sense of belonging in a world that often seems indifferent, if not hostile.

Gang wars, replete with their brutalities, are reminiscent of historical battles fought for territory, power, and economic supremacy. These wars, while seemingly senseless to outsiders, are governed by the same principles that have directed human conflict for eons.

1. **Territory:** Like nations vying for land and resources, gangs often clash over control of areas with strategic or economic value. These 'territories' become centers for illegal trade, providing revenue streams essential for the gang's sustenance.

2. **Power:** Power dynamics within the urban jungle are fluid. As gangs vie for dominance, skirmishes, and wars become inevitable as each group attempts to assert its influence over others.

3. **Economic Gain:** The underground economy, whether it revolves around drugs, illegal weapons, or other contraband, is lucrative. Control over these markets can mean the difference between affluence and destitution for gang members.

4. **Revenge:** In a world where respect is paramount, slights, whether real or perceived, can lead to violent confrontations. Revenge becomes both a matter of pride and a deterrent against future aggression.

5. **Defense:** Defense is intrinsic to survival. For gangs, defending their members, territory, and interests against rival factions is non-negotiable.

Through the lens of survival, actions that might seem immoral or extreme to the average citizen can appear justified within the gang context. Extreme violence, then, becomes a tool, wielded to ensure the gang's continued existence and the safety of its members.

Political Malfeasance:

The Murky Waters of Power and Corruption

"Power tends to corrupt; absolute power corrupts absolutely." - These words by English historian Lord Acton resonate deeply in our understanding of the political arena, no matter the epoch or nation under scrutiny. They succinctly capture the essence of what lies beneath the facade of governance – an underbelly riddled with misuse of power, deceit, and corruption.

Taking the American context as a prototype, the annals of its political history are marred by countless forms of misconduct. While holding significant federal positions, politicians have been convicted of many crimes, ranging from bribery, land fraud, and corruption to more personal felonies like child pornography, possession of cocaine, and sexual misconduct. Others include salary fraud, illegal campaign contributions, kickbacks, obstruction of justice, money laundering, and racketeering.

These instances paint a bleak picture of the very individuals entrusted with shaping a nation's future. It manifests what happens when the intoxicating blend of power, fame, and greed supersedes the mandate of serving the public.

Collage of Totalitarian Leaders. (Each row - left to right) Joseph Stalin, Adolf Hitler, Augusto Pinochet, Mao Zedong, Benito Mussolini, and Kim Il-sung.Wikipedia. Creative Commons Attribution-Share Alike 4.0. VectorVoyager.

Politics and Morality: An Age-old Tussle

The world of politics is paradoxical. While it's rooted in the nobility of public service, it frequently grapples with issues of morality. Like religion, politics also boasts an inherent potential to drive individuals toward extreme behavior. Political ideology can be so deeply ingrained that any difference is perceived as a direct assault on one's beliefs, leading to vehement reactions ranging from violent protests to character assassinations.

Why does this happen?

At its core, the realm of politics influences almost every aspect of our lives. The essential debate revolves around the degree of government intervention necessary to ensure societal order, justice, and well-being.

The Dichotomy of Governance: Minimalist vs. Interventionist

1. The Minimalist Approach: Advocates of limited government intervention argue for a system where governance intrudes as little as possible in citizens' lives. The ethos of this perspective is that justice and opportunity are best served when individuals possess maximum freedom. Here, minimal rules, individual autonomy, and personal freedoms are paramount.

2. The Interventionist Approach: Those favoring a more expansive role in government believe that a larger governing body is imperative to ensure equitable prosperity and justice. This perspective views the government as the pivotal force in instilling fairness and social justice in society.

These polar ideologies lay the groundwork for political clashes. Each side is convinced of the righteousness of its beliefs, leading to actions and reactions that often transgress moral boundaries.

A significant consequence of this ideological warfare is again the demonization of the opposition. The political arena morphs into a battleground, with each side painting the other as the "enemy." When members of an opposing faction are dehumanized, actions against them – no matter how egregious – are easily justified. This rationalization bears an uncanny resemblance to the gang mentality, where the survival and dominance of one's group take precedence over all else.

This culture of vilifying opposition, coupled with the amplification of fake news, threats, and censorship, is perilous. It erodes the very foundation of democratic governance, where dialogue, discussion, and consensus should ideally hold sway.

The political milieu is often governed by the same flawed survival instinct that fuels aggression– the drive to sustain power, even if it means forsaking ethical considerations. The parallels between this and the inner-city gang mentality are striking. Just as gangs prioritize their group's survival over societal norms, political factions, too, at times, seem to place party loyalty and power preservation above national interest.

Dehumanization: An Unraveling of the Human Psyche

The act of seeing another human being as less than human strips away the veneer of civility and decency that we, as a society, strive to cultivate. David Livingstone's "On Inhumanity" probes the dark crevices of this mindset. He posits that dehumanization isn't simply about relegating others to the status of inferiority but is an intricate web that supports oppression and systematic violence. Delving deeper into this concept unveils the profound implications of dehumanization on society and how it might very well be a survival mechanism, albeit a distorted one.

To grasp the nuances of dehumanization, it is vital first to understand its core. According to Livingstone, dehumanization can be encapsulated in the idea of seeing others as less than human. This seemingly straightforward assertion has vast implications. Considering another as not fully human diminishes the moral restraint that would otherwise prevent one from inflicting harm.

In this light, dehumanization is not merely an ideological stance; it's a psychological mechanism. It desensitizes the dehumanizer, rendering them capable of oppressive or violent actions without the accompanying weight of guilt or moral repugnance. As a result, the threshold for cruelty lowers, and acts that would ordinarily be unthinkable become not just conceivable but even justifiable.

While dehumanization is commonly linked to overt acts of violence, Livingstone reminds us that its shadow stretches further, often leading to covert forms of oppression. Consider the caste systems, racial segregation, and discriminatory policies ingrained in various societies. These are not always marked by physical violence, but they're rooted in a foundational belief that certain groups are inherently inferior or less deserving than others.

Such oppressive systems are sustained and legitimized by dehumanizing narratives. They create societal structures that limit access, suppress voices, and perpetuate inequality – all under the guise of preserving social order or upholding cultural values.

Livingstone says dehumanization springs from a desire to inflict harm upon others. This desire could be sparked by various triggers: fear of the unknown, competition for resources, perceived threats to one's supremacy, or deep-seated prejudices passed down through generations.

But why would any individual or group have an intrinsic desire to harm another? One possible explanation lies again in evolution. If we consider our primal instincts, the urge to dominate, eliminate threats, or establish territory has always been essential for survival. Today, in a world governed by socio-political complexities, these instincts don't manifest in hunting or territorial conquests. Instead, they metamorphose into ideologies, policies, and systemic oppressions.

Dehumanization, in this context, acts as a facilitator. It provides an emotional detachment that allows individuals to act on these harmful instincts without the constraint of empathy or moral trepidation.

Livingstone argues that dehumanization serves to disinhibit our most sinister impulses. By viewing others as less than human, the internal moral compass that would ordinarily guide behavior becomes distorted. Acts of cruelty, oppression, and violence are no longer at odds with one's self-perception as a “good” or “moral” being. The dehumanized are perceived as deserving of maltreatment or simply not worthy of the compassion typically extended to fellow humans.

This function of dehumanization can be seen starkly in instances of genocide, where entire groups are exterminated based on a concocted narrative of their "lesser" value. But it also permeates more subtle, everyday biases and prejudices that individuals might harbor.

Preservation of the Species: A Twisted Justification

It's a chilling thought, but from a skewed perspective, dehumanization could be seen as a mechanism for preserving the purity of one's group. By ensuring that one's in-group thrives (even at the cost of another's suffering), individuals might believe they are bolstering the survival chances of their progeny.

However, this rationale overlooks a crucial facet of our evolution: humanity's success has been largely due to collaboration, empathy, and social bonding. While competition and dominance have played roles in our history, it's our capacity for cooperation and understanding that sets us apart.

The implications of Livingstone's exploration into dehumanization are vast. It serves as a poignant reminder of the depths to which humanity can sink when empathy is suspended. However, there lies hope in recognizing and understanding the mechanisms of dehumanization.

By actively promoting empathy, encouraging intergroup dialogue, and challenging dehumanizing narratives, societies can take strides toward creating an environment where every individual is seen, valued, and acknowledged in their full humanity. Recognizing our shared human experience is the first step in bridging divides and building a more inclusive and compassionate world.

> *"There are no nations! There is only humanity. And if we don't come to understand that right soon, there will be no nations, because there will be no humanity."* Isaac Asimov

PART III:
Man's Super Brain As the Enabler

A Deep Dive into the Gray World of Morality

The relationship between morality and human cognition, with its mesmerizingly diverse moves, has been a puzzle and a fascination for millennia. At the heart of this entanglement lies the human brain - a prodigious orchestrator of beliefs, actions, and justifications. The very premise that evil often finds its way under the banner of perceived righteousness brings forth the immense power and susceptibility of the human mind.

In many historical and contemporary scenarios, actions perceived by one group as fundamentally righteous are seen by others as profoundly evil. The Holocaust, 9/11 attacks, and countless other incidents don't originate from a space of clear-cut malevolence. Instead, they're driven by a perceived higher purpose or a more significant good. As echoed in Lucas' "Star Wars Trilogy," Joseph Campbell's notion posits this battle of duality as a timeless struggle of good vs. evil. But is this battle truly black and white?

Morality: The Fluid Benchmark

Morality, at its core, is profoundly subjective. The absence of a universally accepted moral yardstick gives rise to diverse interpretations of right and wrong, further complicated by cultural conditioning. For instance, Solzhenitsyn's reflections in "The Gulag Archipelago" bring forth the disturbing realization that the seeds of both good and evil reside within every individual, challenging simplistic binaries.

Our evolutionary journey has endowed us with a malleable brain, capable of adapting to varied environmental stimuli. Such adaptability extends to our moral compass. "Immorality East and West" offers intriguing insights into these cultural disparities in moral perception. Researchers Emma Buchtel et al. found vast differences in the perception of right and wrong between people living in the United States and those living in China. Here are just a few of their findings:

1. To kill a person:
   90% in the US perceived this to be immoral, but only 57% in China.

2. To intentionally cause harm to someone for your own gain:
   89% in the US perceived this to be immoral, but only 65% in China.

3. To intentionally hurt another person:
   81% in the US perceived this to be immoral, but only 56% in China.

4. To spit on the public street:
   7% in the US perceived this to be immoral, but 70% in China.

5. To be disrespectful to your parents:
   24% in the US, but 76% in China.

6. To talk and laugh loudly in a public place:
   5% in the US perceived this to be immoral, but 44% in China.

What stands out is not the difference in numbers but the underlying cultural narratives that shape these numbers. This spectrum of moral perceptions, molded by cultural nuances, underscores that our inherent predispositions can be profoundly altered based on our upbringing.

For Homo sapiens, this malleability is a double-edged sword. On one hand, it allows for adaptability and evolution. On the other, it implies that beliefs about right and wrong can be easily molded, manipulated, and even weaponized.

Defensive Mechanisms of the Brain

Recognizing oneself as 'evil' destabilizes our psychological equilibrium. Such an acknowledgment threatens our evolved survival mechanisms. Negative self-perceptions, if unchecked, could lead to a cascade of harmful emotional states. But the human brain, ever the protector, deploys a plethora of defensive strategies to maintain self-worth. These strategies, while preserving one's self-image, often blur the lines of goodness and evil.

A pertinent example lies in wartime atrocities. In such contexts, enemy dehumanization becomes a cognitive strategy, enabling soldiers to carry out orders that may otherwise be morally reprehensible to them. The brain, in such instances, recalibrates moral standards to fit the prevailing narrative.

Dan Ariely's "fudge factor" introduces another dimension to this discussion. Ariely says individuals continuously test their moral boundaries, often nudging them for personal gains. Herein lies the conundrum: how much moral compromise is acceptable before one's self-image crumbles?

The human brain's astonishing ability to rationalize actions, even ones that contradict one's moral standards, attests to its incredible flexibility. However, this also poses a grave danger. The line separating minor moral indiscretions from grievous acts becomes porous, allowing individuals to slide down a slippery slope of escalating moral compromises.

Conclusion:

Our advanced neuroprocessor, while being our greatest asset, can also be our Achilles heel when it comes to moral judgment. It allows for adaptability but also opens doors to heinous manipulation. Recognizing the influence of cultural conditioning, evolutionary predispositions, and the brain's defense mechanisms can provide a clearer understanding of morality's fluidity.

The challenge is not to establish a universal moral benchmark but to cultivate a heightened awareness of these influences. Such awareness, combined with continuous introspection, can pave the way for more informed and holistic moral judgments. As we grapple with the eternal battle between right and wrong, it is crucial to remember that this battle is not just external but rages within, influenced significantly by the stories our super brains tell us.

Here are some examples, starting with minor moral infractions to unthinkable ones:

Moral Breach	**Fiction-Making Brain Protecting You**
1. I occasionally lie.	"Everyone lies. I generally lie to protect people from hurting."
2. I selectively run red lights.	"Why stop when no one is there? It's simply a waste of time and gas. I never run red lights when I know cameras are recording violations. I can outsmart the system, and I feel smarter for doing it."
3. I ran a red light that caused an accident but convinced the responding police officer that it was the other person's fault. I did not get a ticket, and she did.	"I can't afford to lose my license and can't afford the fine. I have to drive my kids to school. The officer was simply a dumbhead."
4. I cheated on an exam and got a high grade.	"She uses the same stupid test every year. Everyone cheats. I hate her class. She is an idiot!"
5. I occasionally drink and drive.	"Everyone my age drinks too much. I've never been pulled over. I can handle drinking and driving better than most."
6. I stole some merchandise from a store.	"I need the things I took, and the store will never miss them. I got fired from the store 1 year ago, and I hate the owners. They never paid me what I was worth."
7. I joined a gang that sells drugs on the street.	"I have no choice. Without the gang I have no protection on the street. I joined to survive."

8. I bully some kids at school.	"I hate rich kids. Whenever I can, I make them miserable. I love to see the fear in their eyes."
9. I helped beat a rival gang member to death.	"The rival gang killed one of my best friends. I would kill again if I had the chance. He didn't deserve to live.
10. I cheated on my spouse multiple times.	"We should have never gotten married. We stay together for the sake of the kids, Nothing more. I suspect my wife is doing the same thing."
11. I joined an anti-government protest group. I have no problem committing acts of violence against those who disagree.	"Violence is the best way of garnering attention to an issue. I'm fighting for the oppressed."
12. I have taken bribes as a political figure.	"Everyone does it. It's part of the world I'm in. The money has helped my family in countless ways. I have 4 kids in college."
13. I sometimes vote for things I don't agree with.	"If I don't play ball with my party, I'm gone. Being forced out of Congress would put my family at serious financial risk."
14. I sell illegal drugs on the street that occasionally kill people.	"It's how I survive and take care of my family. I'm only selling them what they want. If it kills them, it's not my fault. They asked for it."
15. I cooperated in the extermination of countless Jews.	"Jews are not humans. They are swine. We were just doing everyone a big favor by eliminating them. If I hadn't cooperated, I would have been killed also."
16. I've sexually abused children.	"I never hurt children, I love them. I would never do harm to a child."

17. I nearly killed some dude for his sneakers.	"I deserve them as much as he does. Hey, it's a rough world out there. I take what I want, particularly from trash."
18. I'm guilty of spouse abuse.	"I watched my father beat the hell out of my mother countless times. I think it's in our genes. My wife knows my triggers but goes right into them anyway. Sometimes, she gets what she deserves."
19. I accidentally cut some guy off in traffic, and he pulled alongside me and fired a round through my window. The bullet barely missed me. I retrieved my gun from the glove compartment, chased him down, and shot him twice.	"The guy was crazy as a loon and deserves to be killed. I did the world a favor!"
20. I ordered our military to invade a neighboring country to enhance the survival of our nation. Thousands of people died on both sides.	"Without increased energy production, better agricultural yields, and more water, our people will not survive. I did it to save our people."

Translated from the Latin, Homo sapiens means ***Wise man.*** After examining the history of humankind through the lens of its moral flaw, perhaps a more apt name might be Homo callidus meaning ***Cunning man***.

The Tug-of-War Between Altruism and Inhumanity

Throughout the annals of human history, the canvas of time has been painted with bold strokes of generosity, compassion, and kindness. From the selfless acts of parents towards their offspring to the colossal benevolence of figures like Mother Teresa, humanity has showcased a boundless capacity for goodness. Yet, interwoven with these threads of altruism are the darkened strands of inhumanity documented throughout this book. Despite our capability for profound kindness, the human species has repeatedly been guilty of perpetrating unimaginable cruelty upon its members. This duality presents a complex paradox: how can the same species that birthed the selflessness of Gandhi also be responsible for atrocities that stain our collective past?

Day in and day out, tales of human warmth and compassion light up homes, neighborhoods, and communities. These stories, though often overshadowed by the cacophony of negative news, are the true pulse of human civilization. From a simple gesture of a neighbor helping another with groceries to grand philanthropic endeavors that aim to elevate the quality of life for countless people, the spectrum of human goodness is vast and varied.

However, a closer examination reveals that this benevolence often exhibits patterns. Altruism, though abundant, is frequently directed towards those we identify closely with - be it family, tribe, or nation. Evolutionary biologists argue that this inclination towards 'kin altruism' is hardwired, with humans more predisposed to help those sharing genetic ties. But can genetics alone explain our propensity to reserve our most profound acts of generosity for those within our identified groups?

Humans are inherently tribal. Historically, survival often depended on strong group cohesion and allegiance. Over time, this developed, as has been repeatedly pointed out, into an instinctual bias towards one's in-group. The 'us versus them' mentality has its roots in our primitive need to ensure the survival and success of our tribe against potential external threats.

While this tribal instinct may have served us well in earlier times, its remnants in contemporary society can be divisive. It's worth noting that while many acts of compassion are directed inwards towards those we consider 'our own,' history is also replete with individuals who transcended these barriers. Figures like Martin Luther King, Mother Cabrini, and Mahatma Gandhi showcased a more encompassing brand of altruism, one that was unconfined by boundaries of race, religion, or nationality.

Humanity's journey is one of growth and introspection. While our past is checkered with both commendable altruism and regrettable inhumanity, our future remains unwritten. By acknowledging both facets of our nature and consciously cultivating and championing our benevolent inclinations, we can aspire to a world where acts of kindness and compassion aren't confined by tribal boundaries and where the flawed side of our nature finds fewer shadows in which to hide.

PART IV:
A New Paradigm for Change

Fixing The Flaw Through Neuroscience

The interplay of brilliant achievement and recurring self-inflicted tragedy is a testament to the complexities of the human condition. While the achievements of our species can be hailed as near-miraculous, our inability to fully understand, manage, or even simply accept our own darker impulses stands as a looming human challenge.

At the center of our dual nature is a vital question: If the modern sapiens brain is capable of such extraordinary accomplishments in various fields, why has it been largely ineffectual in taming its own predispositions to hostility and division?

The answer perhaps lies in our evolutionary lineage. The neural architecture and genetic code that facilitated our rise to dominance did not evolve with the current complexities of globalized human society in mind. The landscape of the world has dramatically changed. Our tribes have grown, borders have expanded, and in the digital age, even transcended physical geography. Yet, the deeply ingrained ‘us vs them’ instinct remains, even when it's counterproductive and potentially catastrophic.

Harnessing the Power of Self-domestication

Humans have successfully domesticated various species, reducing their natural aggressive tendencies and molding them to better suit human environments. Domestication of horses, canines, elephants, camels, and even wild hogs (boars) demonstrates that, under highly specific circumstances, the natural genetic instinct to be aggressive and hostile toward those perceived to be different can indeed be repressed and altered.

Domestication has been defined as a sustained multi-generational, mutualistic relationship in which one organism assumes a significant influence over the reproduction and care of another organism. Put another way, domestication represents a permanent genetic modification of a lineage that leads to specific inherited predispositions. Charles Darwin clearly recognized the difference between conscious selective breeding, where certain desirable traits are deliberately targeted as opposed to traits that evolved from natural selection. Evidence from domesticated species provides data for how a reduction in aggression can transform not only temperament but also social/emotional responses, including cooperative/communication skills. For example, surprisingly, domesticated silver foxes, ferrets, and wolves display increased sensitivity to cooperative communication. Domesticated species also tend to show lower testosterone levels which is linked to dominance and aggressive behavior.

The domestication of animals is premised on certain controlled changes in the environment that lead to genetic alterations over time. Similarly, Homo sapiens, armed with self-awareness, can intentionally change their surroundings to influence genetic expression over generations. This process of 'self-domestication' could be the key to attenuating our aggressive impulses.

Photo by chendongshan (AdobeStock). Photo by by famveldman (AdobeStock).

Photo by Rita Kochmarjova (AdobeStock). Photo by cynoclub (AdobeStock).

Early evidence suggests that we may have already commenced this journey. Over the last few centuries, there has been a perceptible reduction in reactive aggression. Through conscious effort and societal structures, we've been able to create environments that favor collaboration over confrontation. But the challenge is far from over. We need to curtail our predisposition towards proactive aggression, the calculated, purposeful aggression directed towards a target. This is the more insidious form, often responsible for wars, genocides, and other large-scale atrocities.

The fields of epigenetics and neuroplasticity offer tantalizing insights into how our environments can reshape genetic expression and neural networks. They suggest that by fostering environments that promote unity, acceptance, and cooperation, we can shift societal norms, leading to lasting genetic and neurological transformations.

Changing the age-old narratives of division requires concerted effort. The first step is a shared recognition of our shared humanity, realizing that the 'tribal' divisions of the past have no place in a world where interconnectivity must reign supreme.

We can override our instinctual biases through targeted social and emotional learning programs, replacing them with values of understanding, compassion, and cooperation. This is a daunting task, one that requires a collective shift in perspective and proactive investment in reconditioning our emotional and cognitive responses.

The potential for Homo sapiens to reshape its destiny is immense. By actively embracing our capacity for self-domestication, we can address the darker facets of our nature. The call to action is clear: We must prioritize the well-being of the collective, recognizing that our shared future hinges on unity, not division. By taking the lessons from our past and combining them with the scientific insights of the present, we can forge a brighter, more harmonious future for all.

The normal aggression that exists between wolves and humans, birds and dogs, tigers and humans and so forth can virtually disappear under the right environmental conditions.

Important Takeaways:

- Gene expression can be intentionally modified through targeted environmental change.
- Modifications in gene expression can be sustained over multiple generations if the modifications are continuously reinforced.
- Homo sapiens' unique capacity for reflective consciousness makes possible the suppression of the survival at all cost coded mandate. Put simply, Homo sapiens can intentionally apply the domestication process to its own species, which might best be called self-domestication.

Available evidence indicates that Homo sapiens has been self-domesticating for three-hundred years, which is one of the most distinctive features of the species. The species has experienced the greatest success in reducing reactive aggression. As defined here, self-domestication is the intentional use of sapiens' advanced brain to reduce the tendency for both reactive and proactive aggression. The training program presented later in this book is such an example. Research shows that tendencies for both proactive and reactive aggression are strongly influenced by developmental experience, particularly early developmental experience. Twin studies indicate that the propensity for both forms of aggression tends to be stable over time but is mediated to some extent by different genes. It's important to note that sapiens have a much higher propensity for proactive aggression than reactive aggression. Because of this and because the potential for harm to others is much greater with proactive aggression, the primary target of this book will be the reduction of sapiens' propensity for proactive aggression.

The Science of Epigenetics

Epigenetics is the study of observable characteristics, known as marks, from the interaction of genes with the environment. The word "epi" in epigenetics implies "on top of" the genetic basis of inheritance. Changes in one's environment do not alter DNA sequence but clearly can affect the regulation of gene expression.

Epigenetic changes in gene expression can last through multiple cell divisions, some for multiple generations. Such changes are controlled through the action of multiple biological processes, including repressor proteins that attach to silence or alter how genes are expressed. What's important here is that epigenetic modifications can be transmitted to the current generation as well as to the organism's offspring through a process called transgenerational epigenetic inheritance.

In practical terms, this means that the epigenetic changes that must occur for mankind to survive are indeed possible. There is hope, provided the new desired behaviors are repeatedly trained.

The remainder of this book will be devoted to detailing a highly specific sample program for overriding the morally flawed survival-at-all-cost mandate.

Targeted Social and Emotional Learning

Sapiens' moral flaw that sanctions the dehumanization of those perceived as different or competing for the same resources manifests itself cognitively, socially, emotionally, and behaviorally.

As we've learned, dehumanization means seeing others as less valuable, inferior, and not fully human. Denigrating or demonizing someone as less than human, as has been repeatedly and painfully detailed earlier, conveniently absolves the person of guilt or self-condemnation.

The evidence is clear and compelling; sapiens' moral flaw breeds aggression, distrust, contempt, disdain, scorn, and repulsion toward outsiders. The training program proposed here aims to suppress and rewire these largely dysfunctional responses and replace them with the counter-balancing feelings and emotions of kindness, sharing, compassion, empathy, caring and cooperation. Decades of research in the area of social/emotional learning and emotional intelligence confirm the following three things:

1. The most fertile period for social and emotional learning is early childhood but important learnings can occur regardless of age with the right training.

2. Emotional responses such as kindness, compassion, and empathy can be thought of as muscles and are strengthened in the same way physical muscles are strengthened: repeated energy investment. Energy can be invested by writing about the emotion, talking about it, thinking about it, visualizing about it, inner voice self-coaching (commonly called self-talk), and acts of doing.
3. Practice in recognizing, understanding, labeling, expressing and regulating emotions (called RULER by researcher Marc Brackett) accelerates emotional learning and competence.

Neuroscience Fundamentals

Before detailing the specifics of the brain retraining and reconditioning program, it should be helpful to review some basic neuro-science related to learning and development.

1. Our brains are a constant work in progress from conception to death. Both genetics and environment continuously shape how our brains and bodies function, how we think, feel and behave. Neuroplasticity is the brain's ability to change and develop in response to the ever-changing environment. It refers to the capacity of our nervous system to modify itself both functionally and structurally in response to experience. The ability of sapiens' brain to change, grow, and reorganize itself throughout the entire span of life opens the door to self-directed, intentional change. Without this capability, modifying the survival at all cost mandate would not be possible.

2. The younger the brain, the more malleable it is to new learning. The optimal period to embed new neural pathways that can modify and shape gene expression is in childhood and adolescence, as they are known periods of sensitive development. As statesman, Frederick Douglass aptly stated, "It's easier to build strong children than repair broken men." The human brain can, however, be retrained at any stage of life. The older the brain, the more training inputs and the longer time will be required to get the desired results.

3. Changing the brain requires targeted sensory input in the form of sight, hearing, touch, smell, and taste (the 5 sensory portals). The most important inputs for modifying genetic expression will be seeing and watching, listening and hearing, public and private speaking, reading and writing, and using imagery and visualization.

4. The greater the frequency and the higher the intensity of targeted sensory input, the more likely it is that the learnings will get traction and the brain will reorganize itself accordingly.

5. Communication between and among neurons is strengthened by repeated energy investment and weakened by lack of investment. Continuously investing in a higher priority than survival, such as caring for others, and intentionally reverting energy away from such a preoccupation can alter brain function and structure. Repeated sensory input, over time, will transform the neoplastic human brain.

6. Feelings and emotions can be trained in the way muscles are trained, with repeated energy investment. The survival of

the fittest and preservation of one's progeny are powerfully embedded in the primitive architecture of emotion. Enduring change requires pervasive neurological adaptations in the way the brain processes incoming data.

7. An important part of training is strengthening the sheath that forms around nerves to protect and preserve signal strength. The substance is called myelin and allows electrical impulses to be transmitted quickly and efficiently. Myelinating new neural networks that support the suppression of the flawed coded mandate by replacing it with kindness and care for all others is the objective.

Addressing the Flaw Requires Three Steps

From more than 35 years of experience and data collection at the Human Performance Institute, I have learned that lasting change requires the following three-step process:

Step 1. A clear picture of the desired endgame. What is the precise mission to be achieved and what is the motivation driving it?

Step 2. A clear picture of what the truth is now. What is the reality of where we are now relative to the intended mission?

Step 3. A clear call to action to close the gap. Specifically, what intentional and purposeful energy investments must be made to complete the mission?

Before applying the 3-step process to sapiens' moral flaw, a practical example might be helpful. Let's apply it to the issue of world hunger.

Step 1. The endgame is the eradication of world hunger. The mission is to provide safe drinking water and food for everyone living on Earth. The motivation driving it is simply humanity's compassion to end the suffering of millions worldwide.

Step 2. The truth is that more than 8 million people die every year from hunger. Poor nutrition and starvation are responsible for the deaths of as many as 3 million children a year. More than 24,000 people die every day from hunger.

Step 3. A call to action worldwide has been made to address the staggering problem. Bringing public awareness to the scope of human suffering has recruited millions of individuals, countries, and entire nations to assist with the mission.

Here are just a few of the programs that have been formed in response to the urgent call for action:

The United Nations World Food Program (WFP)
Bread for the World
UNICEF
Rise Against Hunger
Action Against Hunger
Meals on Wheels
Project Concern
International Alliance to End Hunger
Feed the Children
Why Hunger
Food for the Hungry
Caritas International
World Central Kitchen

The exciting news is the worldwide effort is making real progress in reducing world hunger. The trend is decidedly positive, but much remains to complete the mission. Unfortunately, there is no current unified call to action to address sapiens' aggressive nature that is clearly linked to the horrors of war, slavery, genocide, prejudice, gang warfare, religious conflicts, and much more.

Let's now apply the 3-step process to Homo sapiens' moral flaw:

Step 1. The end game is to replace the obsession with "me" and "mine" to "we" and "us." The mission is to subvert the inherited predisposition to dehumanize those who are "not us" and strengthen the disposition to treat everyone as family. Taking care of our own means taking care of the entirety of our species.

Step 2. The brutal truth about the human cost of Homo sapiens' moral flaw has been detailed in Part II of this book entitled "The Consequences."

Step 3. This publication represents an urgent call to action to address sapiens' aggressive genetic predisposition before time runs out for our species. A 20-week program for parents, teachers, and coaches is presented in the pages that follow to exemplify how programs could be designed to modify the costly flaw.

7 Strategies:

Strengthening New Emotional Responses to Suppress Sapiens' Moral Flaw

The Journey to Internalizing Emotional Messages:

Emotions are powerful drivers of our behavior. They help shape our interactions, influence our decision-making processes, and determine our overall well-being. The importance of training and refining our emotional intelligence has become more significant than ever in our complex world. The process laid out is about not just knowing but deeply internalizing these messages.

Training Process Expanded:

1. Reading the Targeted Emotional Message:

Location Matters: When you place a message in a location frequently visited, like next to your bed or on your study desk, it imprints more vividly in your consciousness.

Frequency: The more you read, the deeper the ingraining. Think of it as repetition building muscle but for your brain.

2. Writing about the Emotional Message:

Longhand Writing: The act of writing longhand connects more parts of the brain than typing. It's therapeutic and reinforces memory.

Reflective Journaling: By continually adding to what you've already written, you build upon previous insights, promoting deeper understanding and personal growth.

3. Talking about the Emotional Message:

Family Discussions: Engaging family members ensures shared understanding and collective emotional growth. It also aids in rooting the message in real-life situations.

Youth Engagement: Young minds are malleable. Discussing these topics with them not only educates but also fosters a more emotionally aware generation.

4. Thinking about the Message:

Personal Relevance: Making the message relevant to your personal life anchors it to real-world scenarios.

Family-Centric Thinking: Considering its application for immediate family makes the message a shared lived experience.

5. Visualization of the message:

Sensory Imagination: The more senses involved in the imagination, the more vivid and memorable the experience. Visualizing acts of kindness can prime one to enact them in reality.

6. Self-coaching about the message:

Power of Voice: There's something incredibly potent about coaching oneself, either privately or publicly. It's the brain coaching itself.

7. Acting on the Message:

Tangible Actions: Actions, they say, speak louder than words. Living the message solidifies its essence in one's mind and serves as a beacon for others.

It's evident that Homo sapiens remarkable brain, shaped by the right messages and neural inputs, can guide the species towards a far more harmonious existence. The following proposed training methods and essential messages are not just strategies but a way of life. When imbibed in our daily routines, they can pave the way for a balanced coexistence, both with ourselves and with others.

Since parents, teachers, and coaches will likely have the greatest opportunity for influence, the following 20 scrolled messages should be helpful in planning specific training strategies.

The 20 Emotional Messages are to be Constantly Uploaded into Sapiens' Neuro-processor Using the 7 Strategies just reviewed.

20 Week Training Program

WEEK 1

You are part of the family of mankind
We are all descendants of the same original parents. We are all family. Differences in skin color, physical appearance, race size, etc. are like ornaments on the tree of mankind to be appreciated and celebrated.
One of our highest values is diversity.

WEEK 1

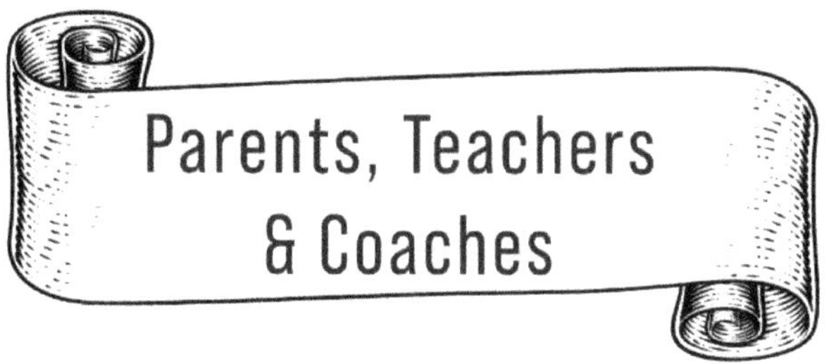

Core message to be conveyed to youth:

You are part of the family of mankind

REFLECTIONS:

1. What does family mean to you?
2. What does the family of mankind signify?
3. If someone is family, do you treat them differently? Are you more forgiving, tolerant, respectful?
4. Why is diversity important?

WEEK 1

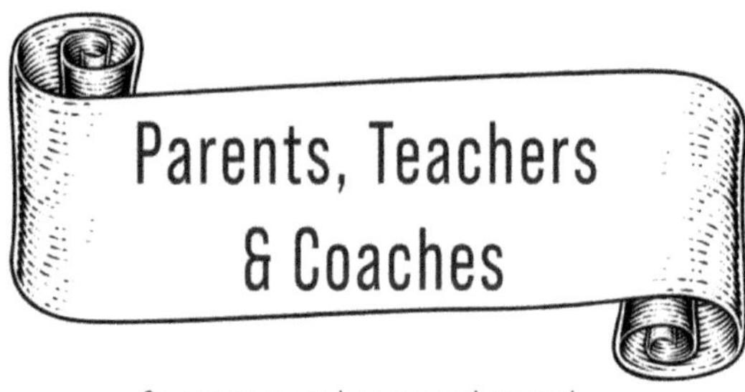

Core message to be conveyed to youth:

You are part of the family of mankind

Explore innovative ways to have your children, students or athletes repeatedly uploald this message by:

1. Having them read about it
2. Having them write about it.
3. Having them talk about it.
4. Having them think about it.
5. Having them visualizing themselves being it.
6. Having them coach themselves with self-talk.
7. Having them act the message out in behavior.

WEEK 2

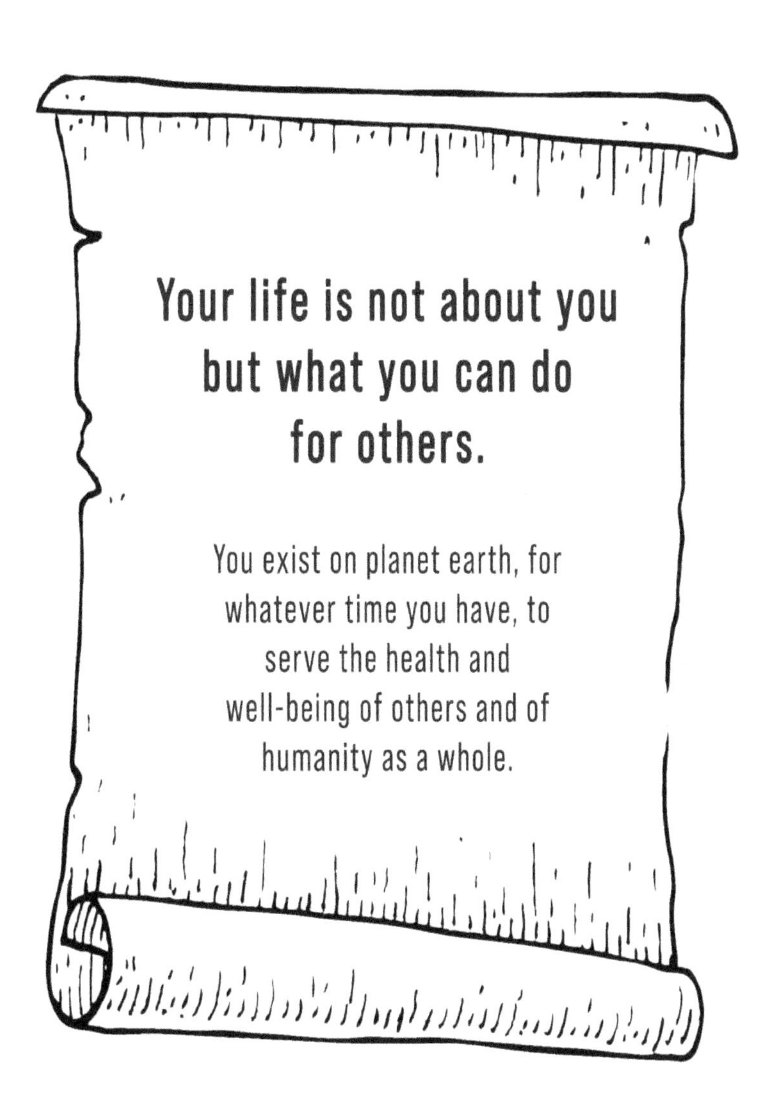
Your life is not about you but what you can do for others.
You exist on planet earth, for whatever time you have, to serve the health and well-being of others and of humanity as a whole.

WEEK 2

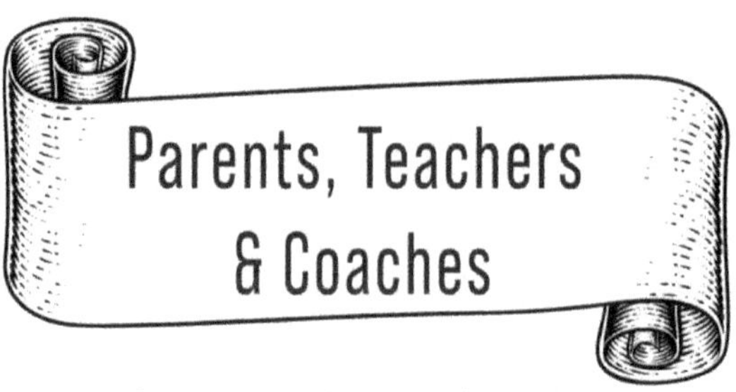

Core message to be conveyed to youth:

Your life is not about you but what you can do for others.

REFLECTIONS:

1. What can you do this week for others?
2. How can you serve the health and well-being of others this week?
3. How do you feel about yourself when you help others?

WEEK 2

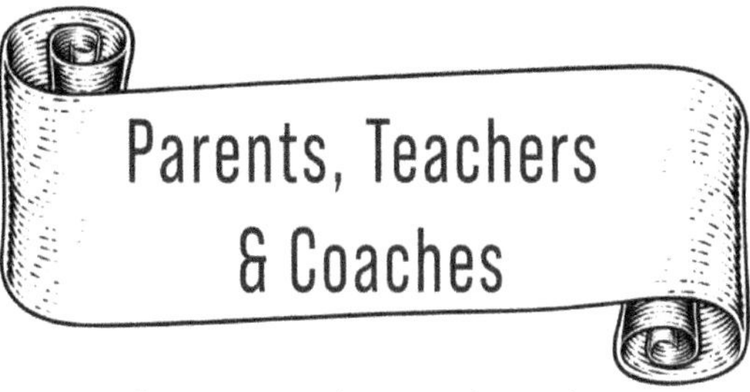

Core message to be conveyed to youth:

Your life is not about you but what you can do for others.

Explore innovative ways to have your children, students or athletes repeatedly uploald this message by:

1. Having them read about it
2. Having them write about it.
3. Having them talk about it.
4. Having them think about it.
5. Having them visualizing themselves being it.
6. Having them coach themselves with self-talk.
7. Having them act the message out in behavior.

WEEK 3

You get a life by giving your life away to a cause bigger than yourself.
Investing in others returns much more than you give.

WEEK 3

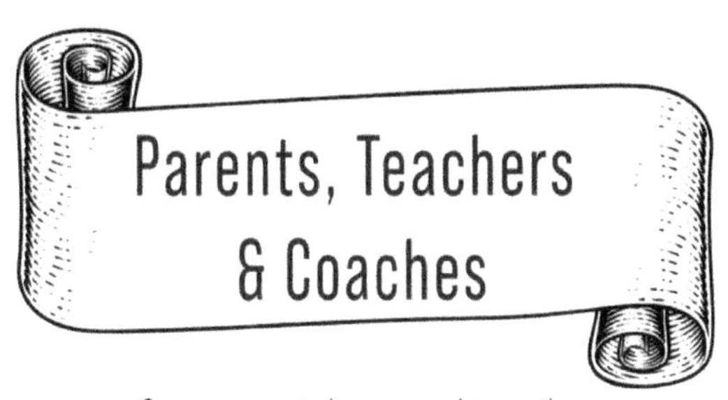

Core message to be conveyed to youth:

You get a life by giving your life away to a cause bigger than yourself.

REFLECTIONS:

1. Give examples of giving your life to a cause bigger than yourself.
2. What do you get back when you give to others?
3. Give an example of what you could do today to align yourself with a cause bigger than yourself.

WEEK 3

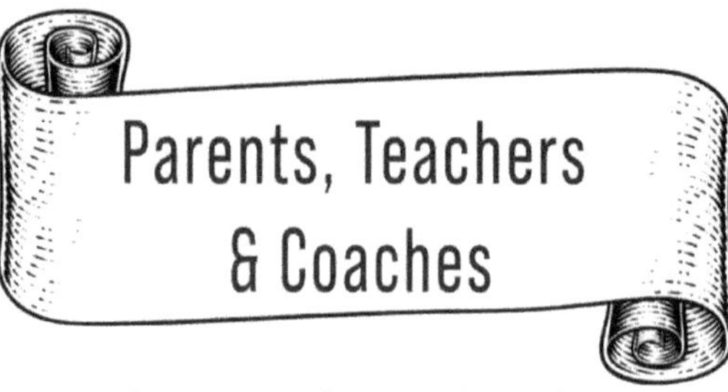

Core message to be conveyed to youth:

You get a life by giving your life away to a cause bigger than yourself.

Explore innovative ways to have your children, students or athletes repeatedly uploald this message by:

1. Having them read about it
2. Having them write about it.
3. Having them talk about it.
4. Having them think about it.
5. Having them visualizing themselves being it.
6. Having them coach themselves with self-talk.
7. Having them act the message out in behavior.

WEEK 4

When you share what you
have with others,
your heart opens up
to goodness.
Be generous!

WEEK 4

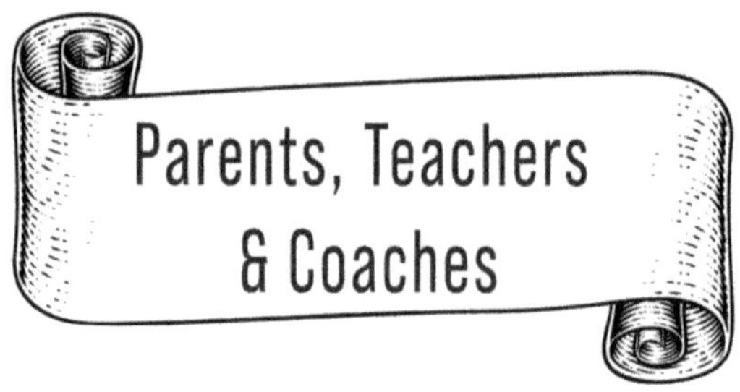

Core message to be conveyed to youth:

When you share what you have with others, your heart opens up to goodness.

REFLECTIONS:

1. Give an example of your sharing in the last week?
2. How do you feel when someone is generous towards you?
3. Give an example of great generosity.

WEEK 4

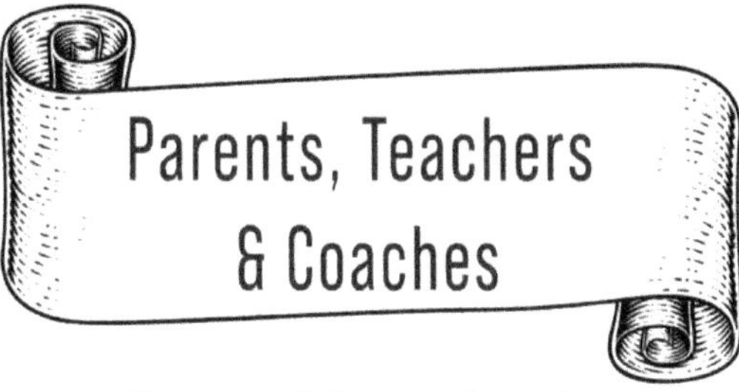

Core message to be conveyed to youth:

When you share what you have with others, your heart opens up to goodness.

Explore innovative ways to have your children, students or athletes repeatedly uploald this message by:

1. Having them read about it
2. Having them write about it.
3. Having them talk about it.
4. Having them think about it.
5. Having them visualizing themselves being it.
6. Having them coach themselves with self-talk.
7. Having them act the message out in behavior.

WEEK 5

Happiness comes from helping others achieve happiness.
Lasting happiness can only come from human connection.

WEEK 5

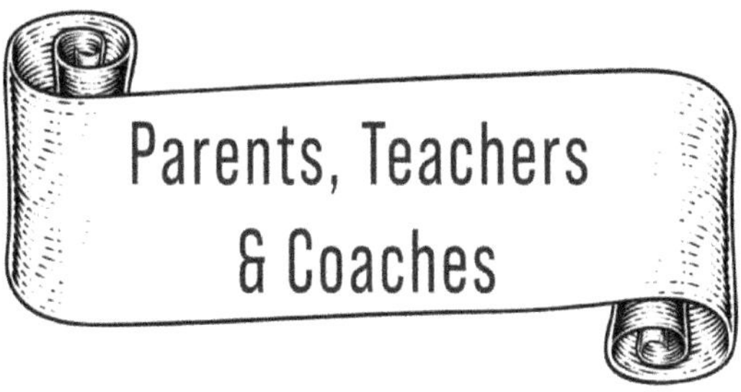

Core message to be conveyed to youth:

Happiness comes from helping others achieve happiness.

REFLECTIONS:

1. How do you feel when you make someone else happy?
2. Describe when you experience the most happiness (when you are the happiest).
3. What does happiness feel like? How is happiness connected to others?

WEEK 5

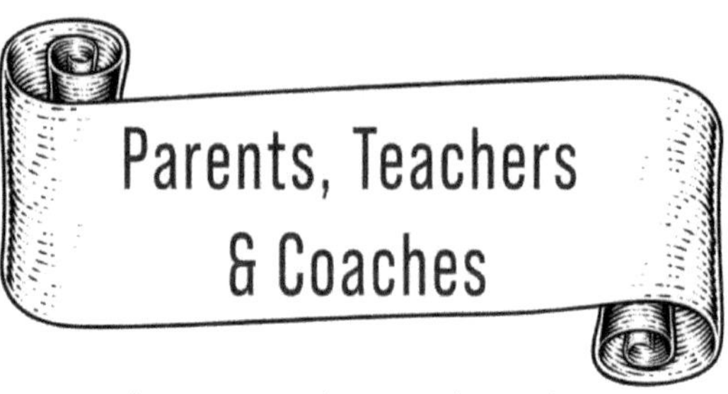

Core message to be conveyed to youth:

Happiness comes from helping others achieve happiness.

Explore innovative ways to have your children, students or athletes repeatedly uploald this message by:

1. Having them read about it
2. Having them write about it.
3. Having them talk about it.
4. Having them think about it.
5. Having them visualizing themselves being it.
6. Having them coach themselves with self-talk.
7. Having them act the message out in behavior.

WEEK 6

Treat others the way you wish to be treated.
Most inhumane acts can be stopped immediately when this message dominates

WEEK 6

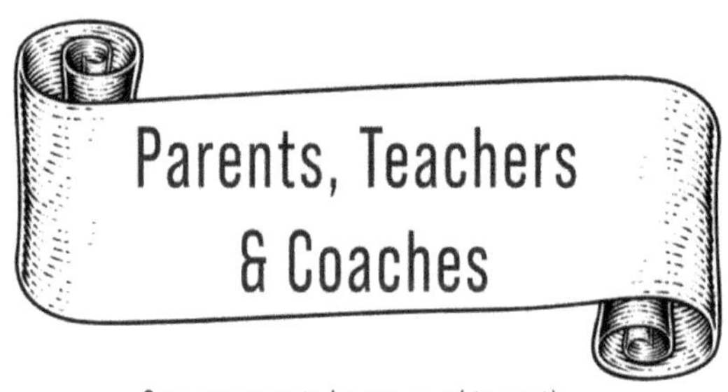

Core message to be conveyed to youth:

Treat others the way you wish to be treated.

REFLECTIONS:

1. Do you ever think about treating others the way you would like to be treated?
2. Give an actual example of when you intentionally treated another person the way you would like to be treated.
3. What if you always use this principle in dealing with others?

WEEK 6

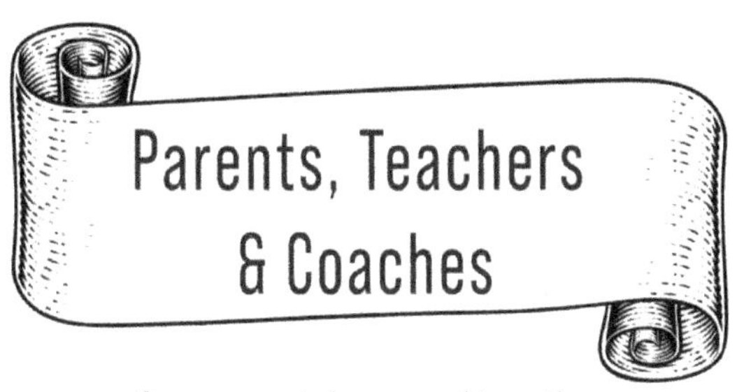

Core message to be conveyed to youth:

Treat others the way you wish to be treated.

Explore innovative ways to have your children, students or athletes repeatedly uploald this message by:

1. Having them read about it
2. Having them write about it.
3. Having them talk about it.
4. Having them think about it.
5. Having them visualizing themselves being it.
6. Having them coach themselves with self-talk.
7. Having them act the message out in behavior.

WEEK 7

You create a more caring, loving, compassionate world by being these in the reality of your own life.
Be the change you want to see in others.

WEEK 7

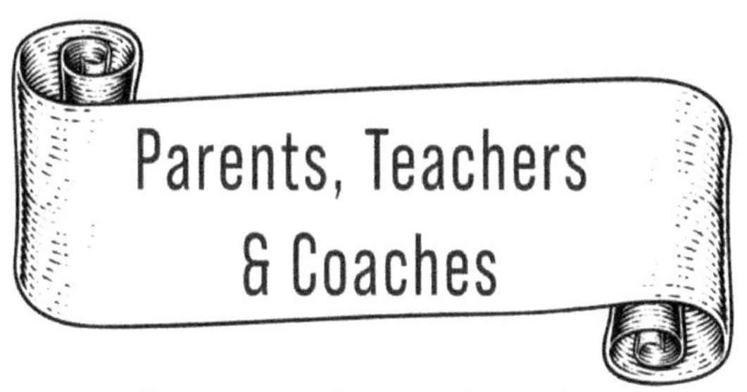

Core message to be conveyed to youth:

You create a more caring, loving, compassionate world by being these in the reality of your own life.

REFLECTIONS:

1. The best way to teach something to others is to be it yourself - explain.
2. Give examples of how this works in your life.
3. How can you be more caring, loving, compassionate?

WEEK 7

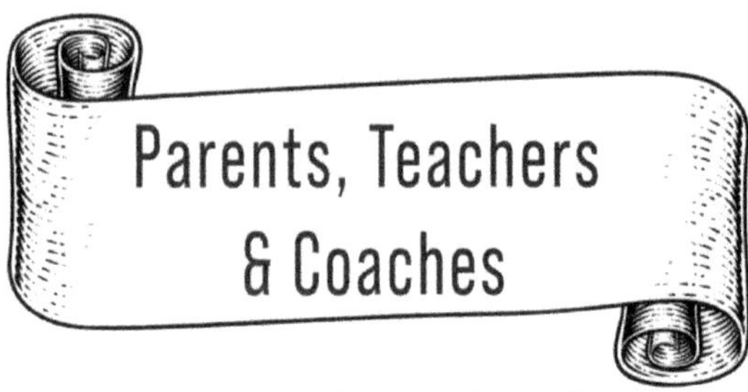

Core message to be conveyed to youth:

You create a more caring, loving, compassionate world by being these in the reality of your own life.

Explore innovative ways to have your children, students or athletes repeatedly uploald this message by:

1. Having them read about it
2. Having them write about it.
3. Having them talk about it.
4. Having them think about it.
5. Having them visualizing themselves being it.
6. Having them coach themselves with self-talk.
7. Having them act the message out in behavior.

WEEK 8

Make kindness your greatest strength.
The foundation of all goodness is kindness!

WEEK 8

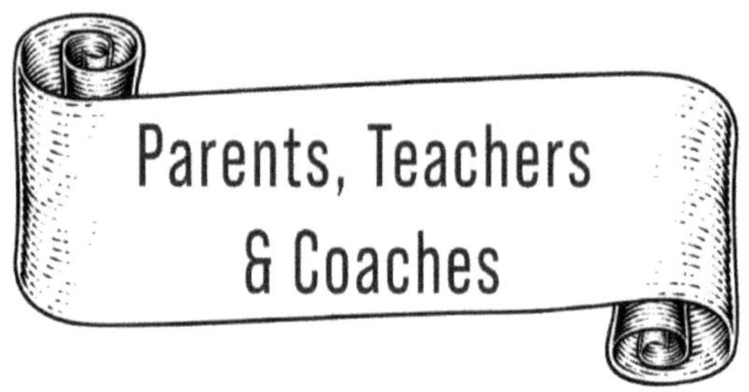

Core message to be conveyed to youth:

Make kindness your greatest strength.

REFLECTIONS:

1. Who is the kindest person in your life?
2. How does that person make you feel?
3. Give examples of how you can show kindness in your everyday life.

WEEK 8

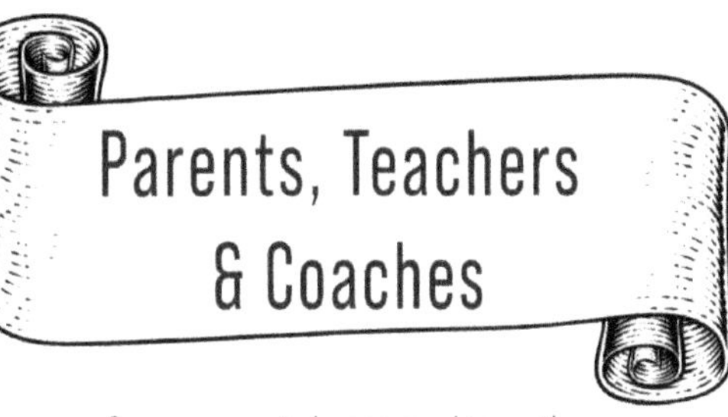

Core message to be conveyed to youth:

Make kindness your greatest strength.

Explore innovative ways to have your children, students or athletes repeatedly uploald this message by:

1. Having them read about it
2. Having them write about it.
3. Having them talk about it.
4. Having them think about it.
5. Having them visualizing themselves being it.
6. Having them coach themselves with self-talk.
7. Having them act the message out in behavior.

WEEK 9

WEEK 9

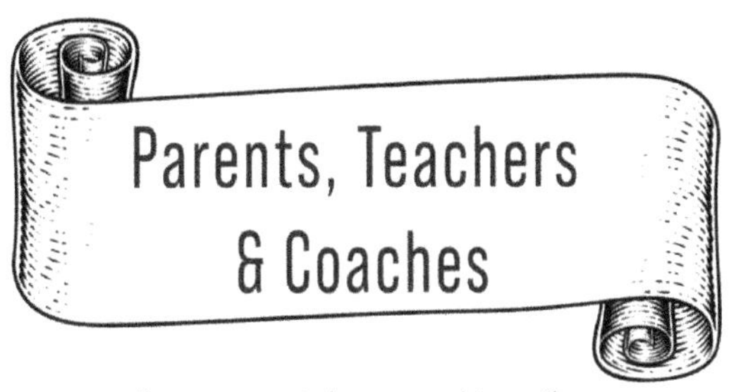

Core message to be conveyed to youth:

There is goodness in everyone.

REFLECTIONS:

1. Think of someone you don't like. Where is goodness?
2. When you can see goodness in someone, how does it change how you feel about them?
3. What is goodness?

WEEK 9

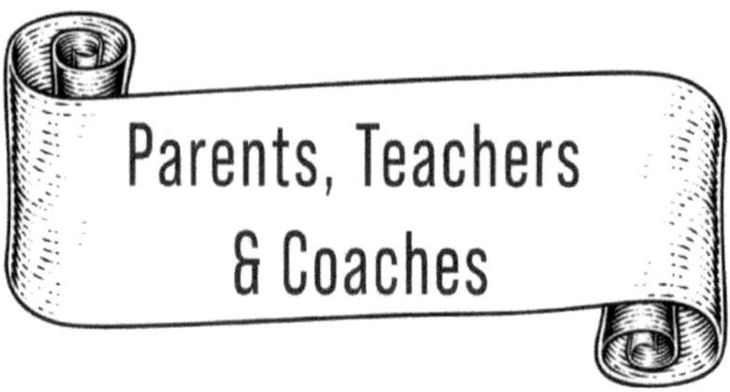

Core message to be conveyed to youth:

There is goodness in everyone.

Explore innovative ways to have your children, students or athletes repeatedly uploald this message by:

1. Having them read about it
2. Having them write about it.
3. Having them talk about it.
4. Having them think about it.
5. Having them visualizing themselves being it.
6. Having them coach themselves with self-talk.
7. Having them act the message out in behavior.

WEEK 10

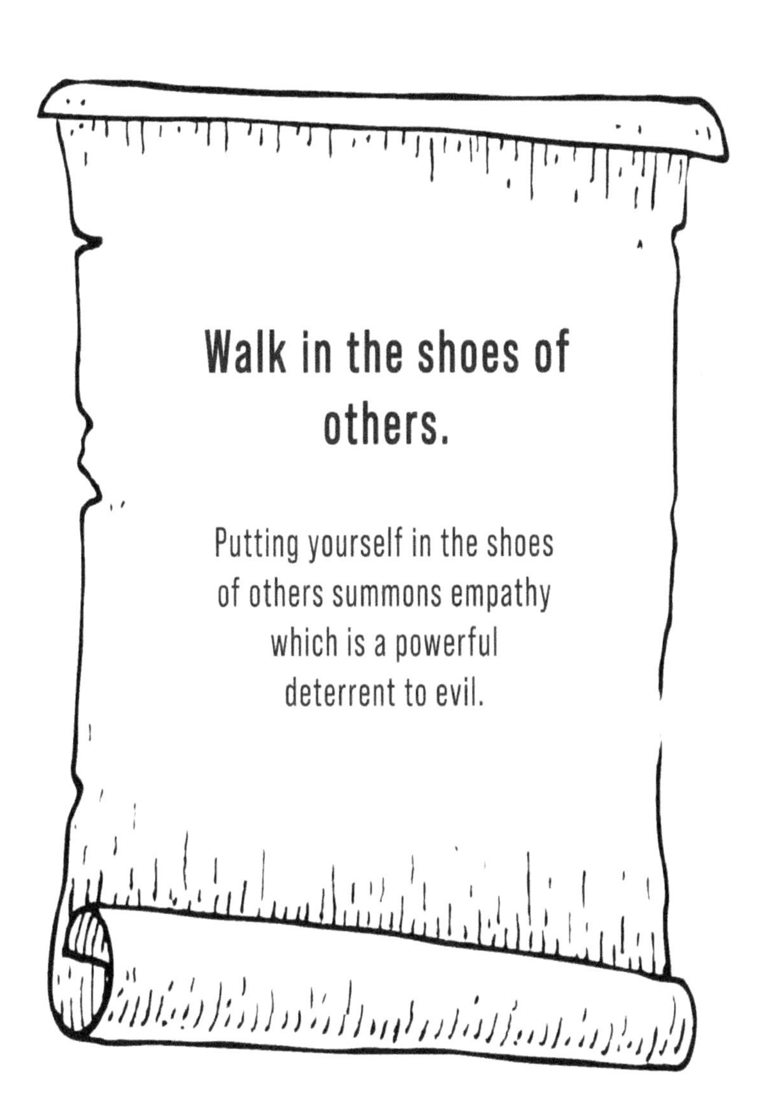
Walk in the shoes of others.
Putting yourself in the shoes of others summons empathy which is a powerful deterrent to evil.

WEEK 10

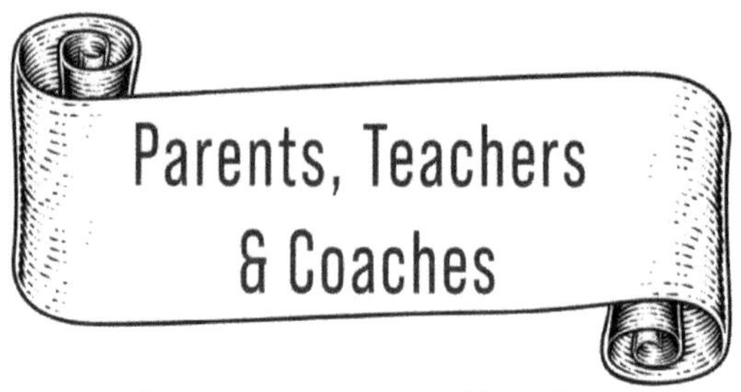

Core message to be conveyed to youth:

Walk in the shoes of others.

REFLECTIONS:

1. What must you do to put yourself in the shoes of others?
2. What's the purpose of doing so?
3. What happens to you when you do step into another person's shoes?

WEEK 10

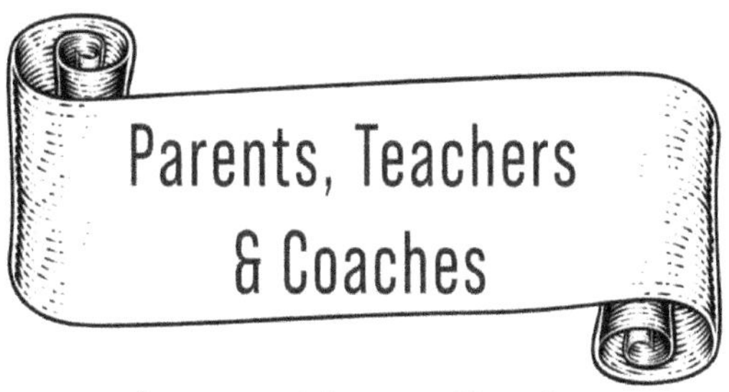

Core message to be conveyed to youth:

Walk in the shoes of others.

Explore innovative ways to have your children, students or athletes repeatedly uploald this message by:

1. Having them read about it
2. Having them write about it.
3. Having them talk about it.
4. Having them think about it.
5. Having them visualizing themselves being it.
6. Having them coach themselves with self-talk.
7. Having them act the message out in behavior.

WEEK 11

WEEK 11

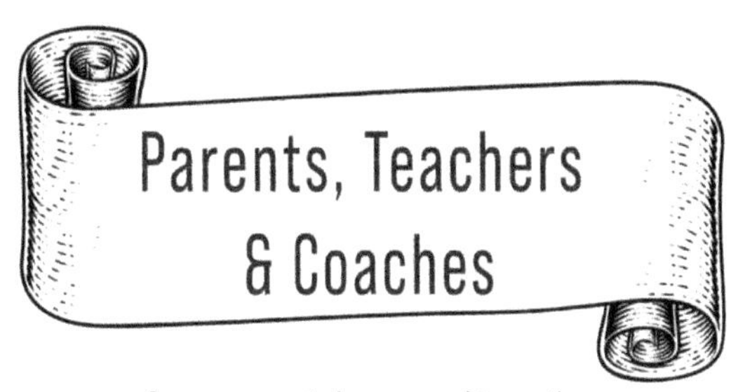

Core message to be conveyed to youth:

Make loving others the most important part of who you are.

REFLECTIONS:

1. What does loving others mean to you?
2. How do you show love to others in your everyday life?
3. Who are the most loving people you know and why?

WEEK 11

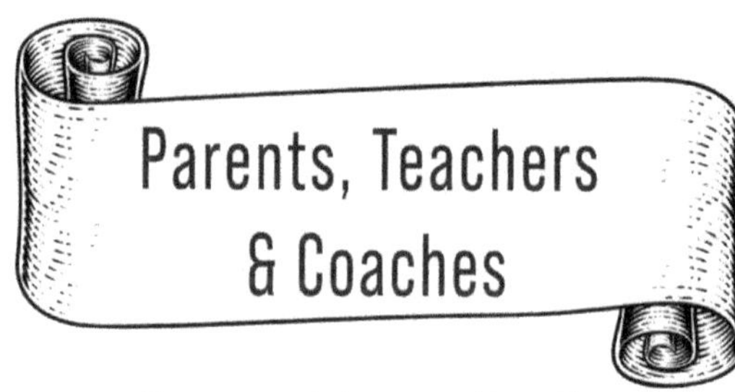

Core message to be conveyed to youth:

Make loving others the most important part of who you are.

Explore innovative ways to have your children, students or athletes repeatedly uploald this message by:

1. Having them read about it
2. Having them write about it.
3. Having them talk about it.
4. Having them think about it.
5. Having them visualizing themselves being it.
6. Having them coach themselves with self-talk.
7. Having them act the message out in behavior.

WEEK 12

Be a loyal, trusted friend
to others and you will be
surrounded by like-
minded friends.
Friendship that is predicated
on helping others forms a
human connection that
both heals and inspires.

WEEK 12

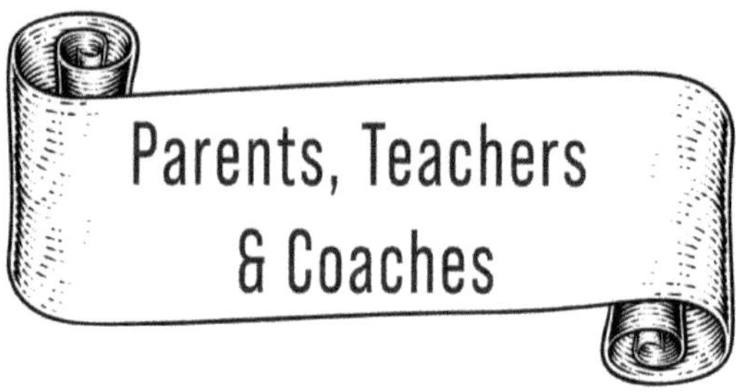

Core message to be conveyed to youth:

Be a loyal, trusted friend to others and you will be surrounded by like-minded friends.

REFLECTIONS:

1. What does a loyal, trusted friend really mean?
2. Give examples of how you have helped your friends in the past.
3. Why are friends important in life?

WEEK 12

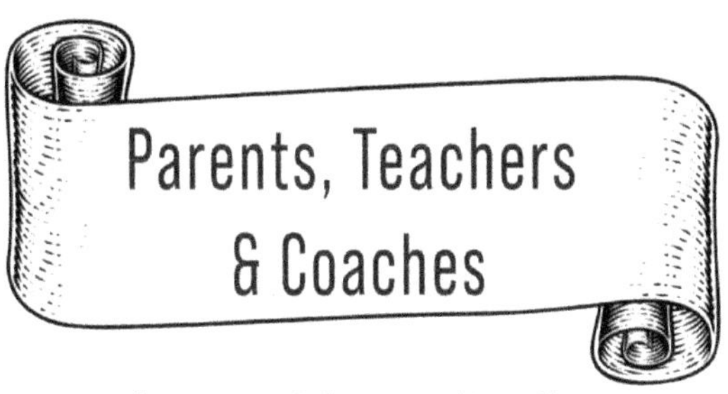

Core message to be conveyed to youth:

Be a loyal, trusted friend to others and you will be surrounded by like-minded friends.

Explore innovative ways to have your children, students or athletes repeatedly uploald this message by:

1. Having them read about it
2. Having them write about it.
3. Having them talk about it.
4. Having them think about it.
5. Having them visualizing themselves being it.
6. Having them coach themselves with self-talk.
7. Having them act the message out in behavior.

WEEK 13

Be good to others!
To be a good person, you have
to do good for others.

WEEK 13

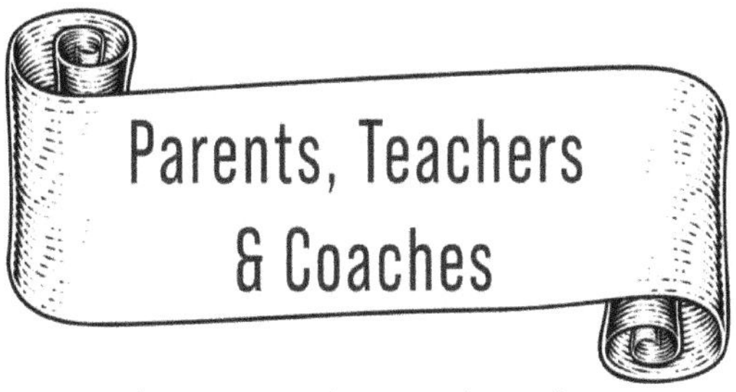

Core message to be conveyed to youth:

Be good to others!

REFLECTIONS:

1. Give examples of being good to others in the last two weeks.
2. Are you a good person? How do you know?
3. How can you improve?

WEEK 13

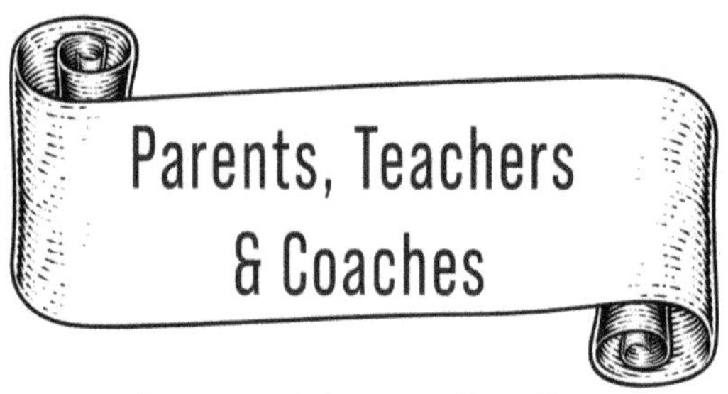

Core message to be conveyed to youth:

Be good to others!

Explore innovative ways to have your children, students or athletes repeatedly uploald this message by:

1. Having them read about it
2. Having them write about it.
3. Having them talk about it.
4. Having them think about it.
5. Having them visualizing themselves being it.
6. Having them coach themselves with self-talk.
7. Having them act the message out in behavior.

WEEK 14

WEEK 14

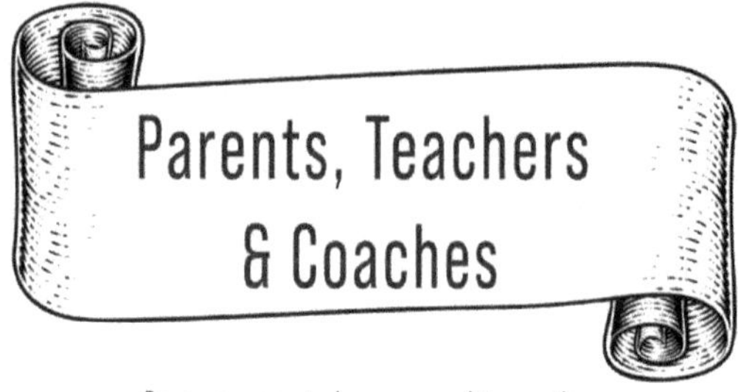

Core message to be conveyed to youth:

Caring and sharing reveal your goodness.

REFLECTIONS:

1. How does one show that he or she cares?
2. Is it more work to care and share or not care and share?
3. Who should you care and share more with?

WEEK 14

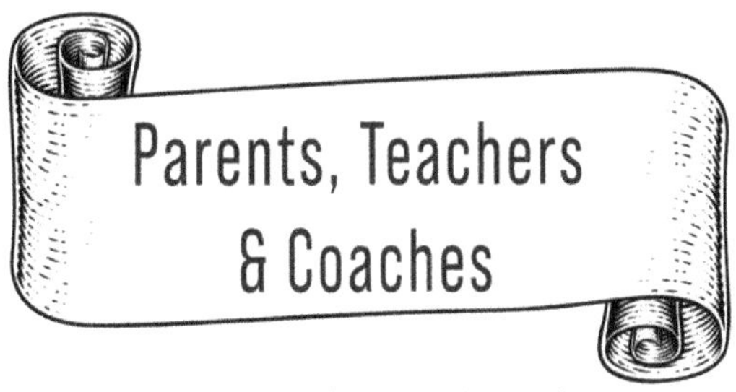

Core message to be conveyed to youth:

Caring and sharing reveal your goodness.

Explore innovative ways to have your children, students or athletes repeatedly uploald this message by:

1. Having them read about it
2. Having them write about it.
3. Having them talk about it.
4. Having them think about it.
5. Having them visualizing themselves being it.
6. Having them coach themselves with self-talk.
7. Having them act the message out in behavior.

WEEK 15

WEEK 15

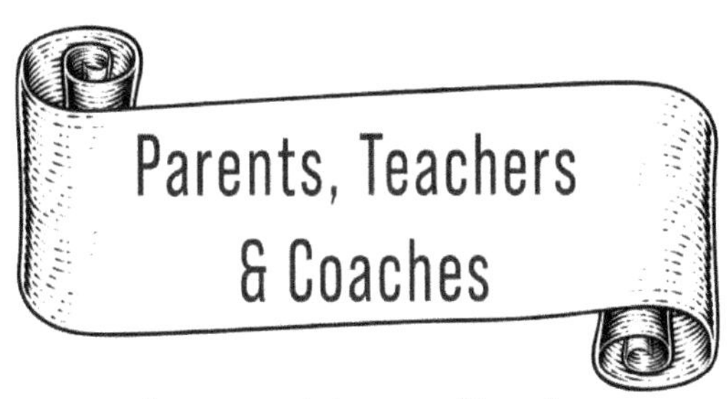

Core message to be conveyed to youth:

Be more of a giver than a taker in life.

REFLECTIONS:

1. Do you think of yourself as more of a giver or a taker in life?
2. Why is this question important?
3. How do you qualify yourself as more of a giver than a taker?

WEEK 15

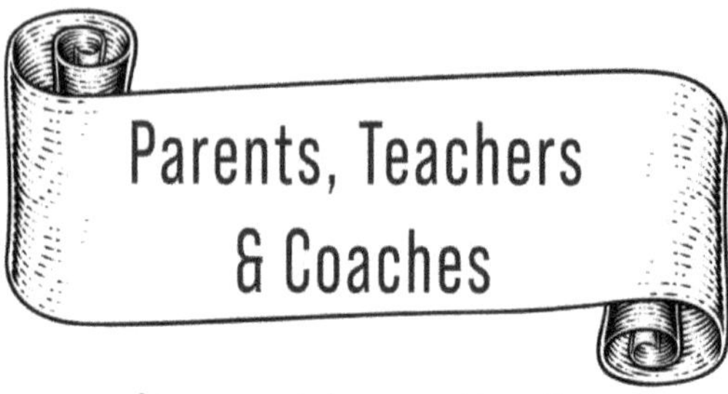

Core message to be conveyed to youth:

Be more of a giver than a taker in life.

Explore innovative ways to have your children, students or athletes repeatedly uploald this message by:

1. Having them read about it
2. Having them write about it.
3. Having them talk about it.
4. Having them think about it.
5. Having them visualizing themselves being it.
6. Having them coach themselves with self-talk.
7. Having them act the message out in behavior.

WEEK 16

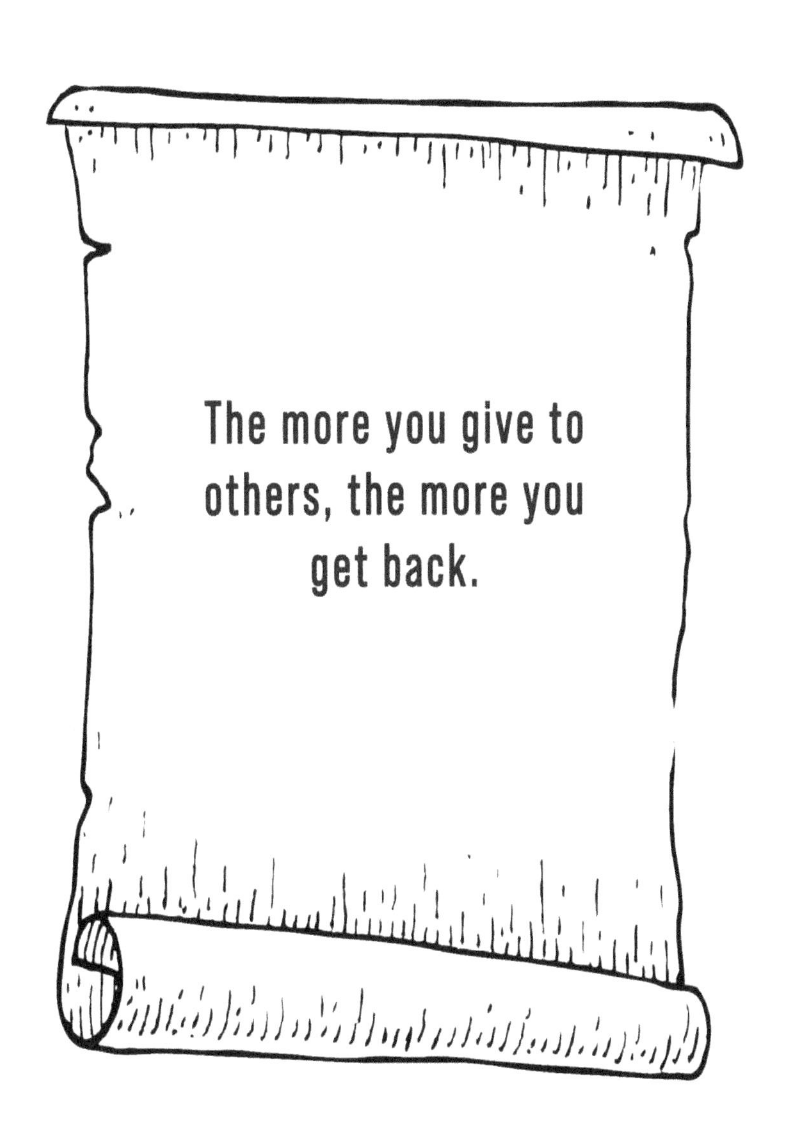

WEEK 16

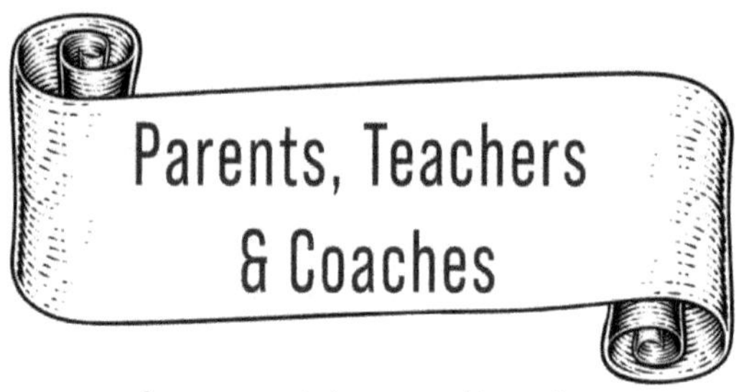

Core message to be conveyed to youth:

The more you give to others, the more you get back.

REFLECTIONS:

1. What do you get when you give to others?
2. Give concrete examples from your life.
3. When is the best time to give to others?

WEEK 16

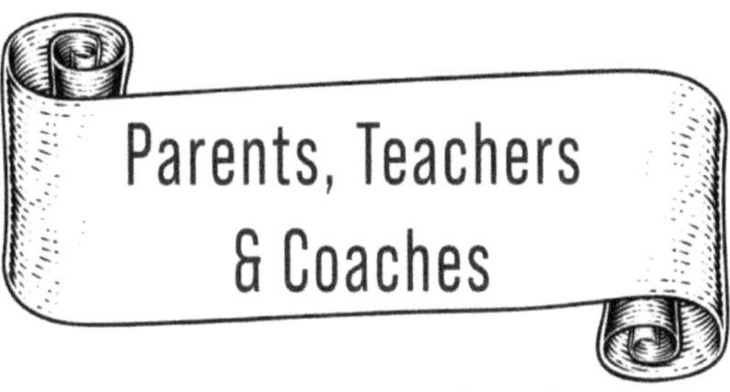

Core message to be conveyed to youth:

The more you give to others, the more you get back.

Explore innovative ways to have your children, students or athletes repeatedly uploald this message by:

1. Having them read about it
2. Having them write about it.
3. Having them talk about it.
4. Having them think about it.
5. Having them visualizing themselves being it.
6. Having them coach themselves with self-talk.
7. Having them act the message out in behavior.

WEEK 17

Be grateful and humble
and the door to personal
fulfillment will
open wide.

WEEK 17

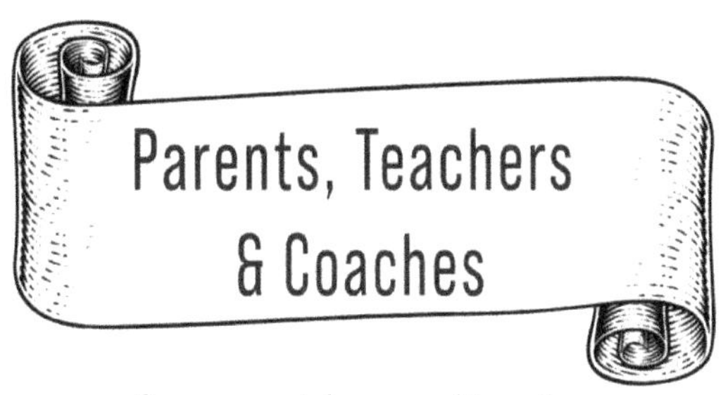

Core message to be conveyed to youth:

Be grateful and humble and the door to personal fulfillment will open wide.

REFLECTIONS:

1. How do you show gratefulness? Are you a grateful person?
2. Are you a humble person? How do you know?
3. Who are the most grateful, humble, people you know?

WEEK 17

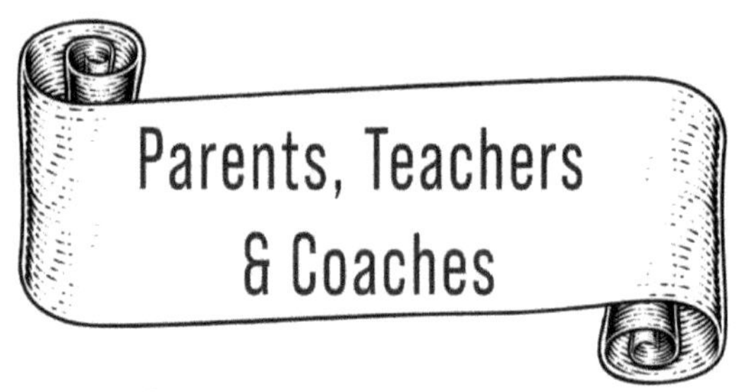

Core message to be conveyed to youth:

Be grateful and humble and the door to personal fulfillment will open wide.

Explore innovative ways to have your children, students or athletes repeatedly uploald this message by:

1. Having them read about it
2. Having them write about it.
3. Having them talk about it.
4. Having them think about it.
5. Having them visualizing themselves being it.
6. Having them coach themselves with self-talk.
7. Having them act the message out in behavior.

WEEK 18

Loving others, all others,
is the only solution to
saving mankind.

WEEK 18

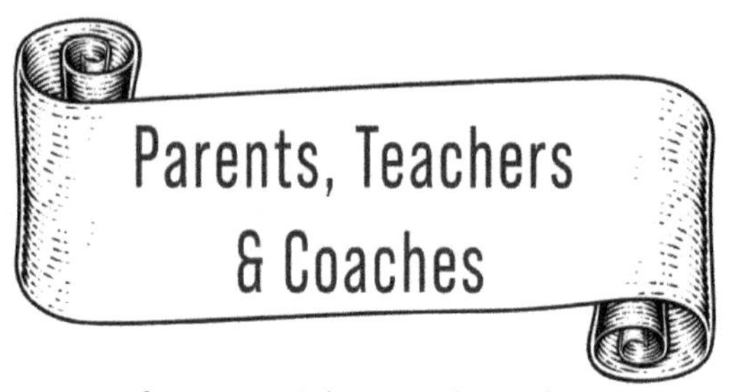

Core message to be conveyed to youth:

Loving others, all others, is the only solution to saving mankind.

REFLECTIONS:

1. In what way might loving others save mankind?
2. What is love?
3. How do you show love for others in your life?

WEEK 18

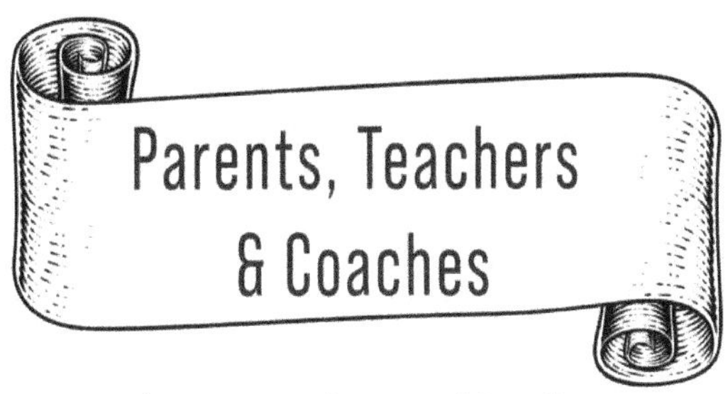

Core message to be conveyed to youth:

Loving others, all others, is the only solution to saving mankind.

Explore innovative ways to have your children, students or athletes repeatedly uploald this message by:

1. Having them read about it
2. Having them write about it.
3. Having them talk about it.
4. Having them think about it.
5. Having them visualizing themselves being it.
6. Having them coach themselves with self-talk.
7. Having them act the message out in behavior.

WEEK 19

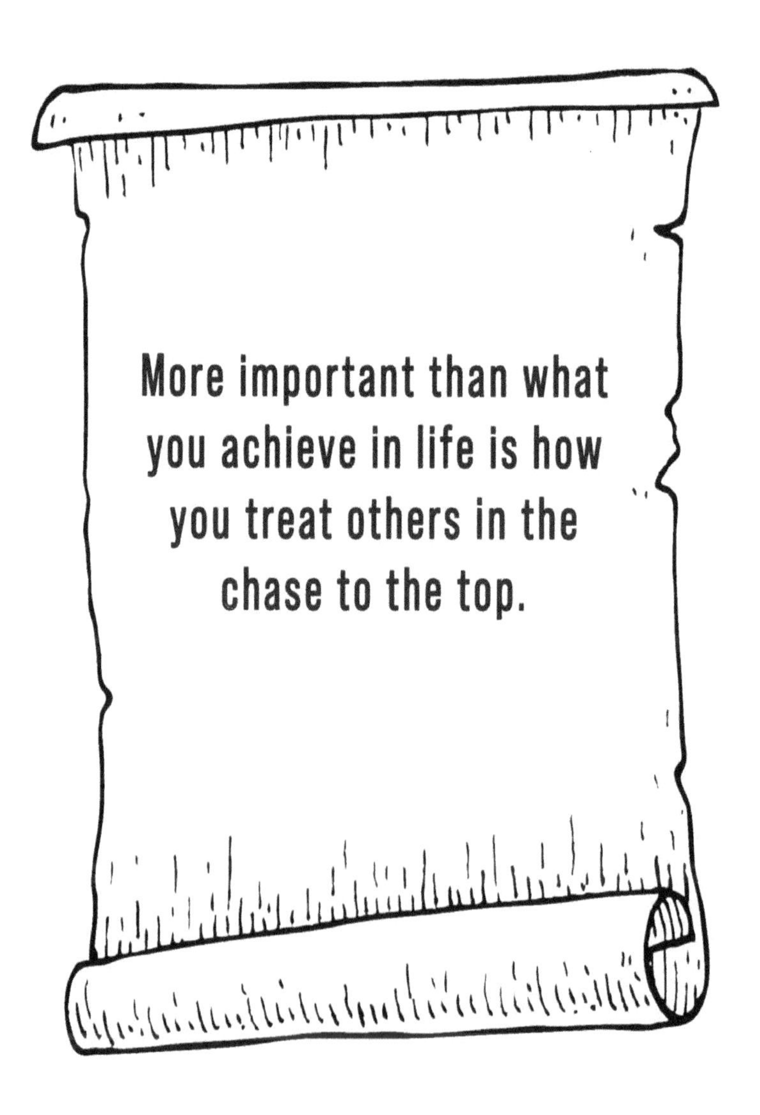
More important than what
you achieve in life is how
you treat others in the
chase to the top.

WEEK 19

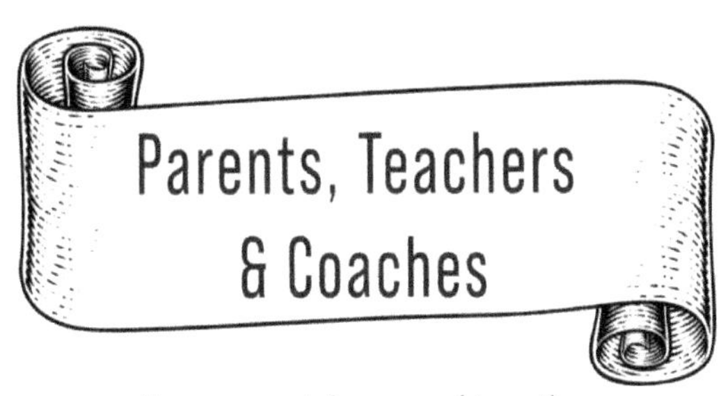

Core message to be conveyed to youth:

More important than what you achieve in life is how you treat others in the chase to the top.

REFLECTIONS:

1. Why is ***how you achieve*** something more important than ***what you achieve*** (how you treat others).
2. Can you achieve great things and treat people well?
3. What if one cheats his/her way to the top?

WEEK 19

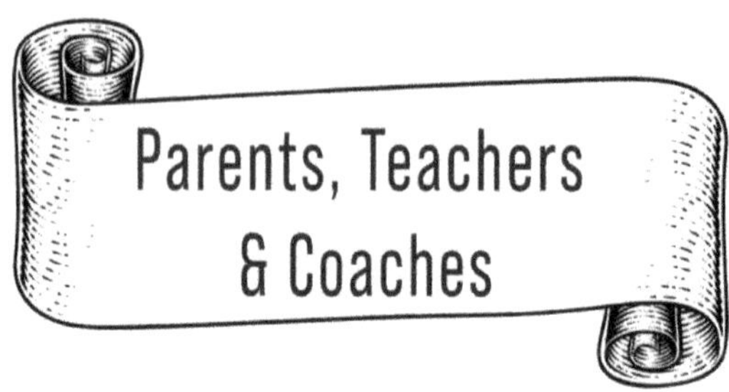

Core message to be conveyed to youth:

More important than what you achieve in life is how you treat others in the chase to the top.

Explore innovative ways to have your children, students or athletes repeatedly uploald this message by:

1. Having them read about it
2. Having them write about it.
3. Having them talk about it.
4. Having them think about it.
5. Having them visualizing themselves being it.
6. Having them coach themselves with self-talk.
7. Having them act the message out in behavior.

WEEK 20

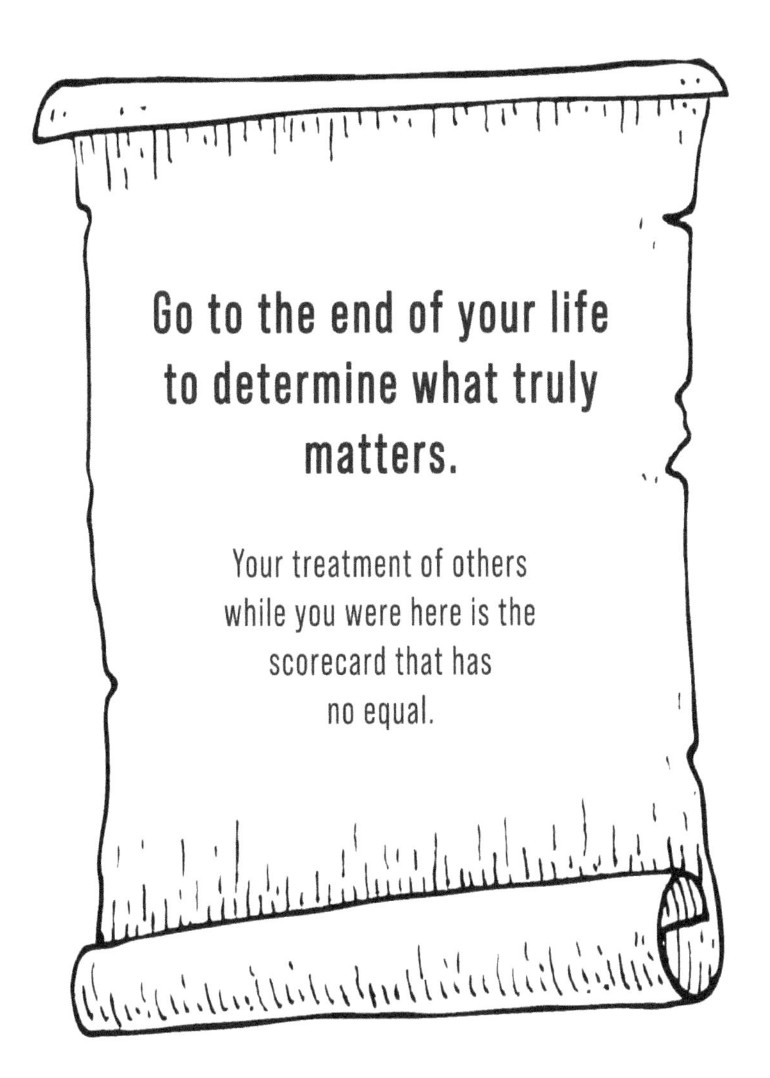
Go to the end of your life to determine what truly matters.
Your treatment of others while you were here is the scorecard that has no equal.

WEEK 20

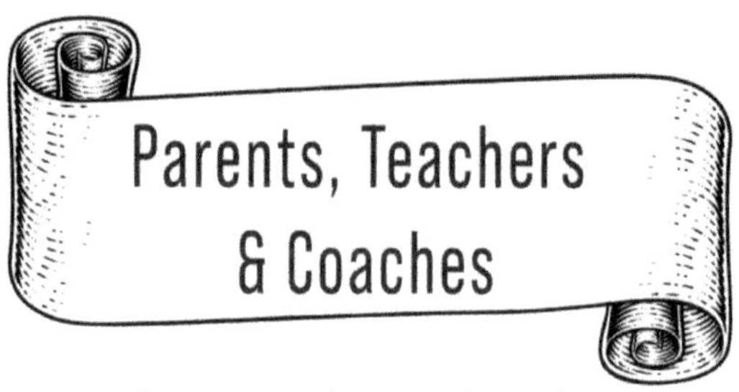

Core message to be conveyed to youth:

Go to the end of your life to determine what truly matters.

REFLECTIONS:

1. What words would you like inscribed on your tombstone that best describes who you most want to be in life?
2. What would you like your eulogy to say about you?

WEEK 20

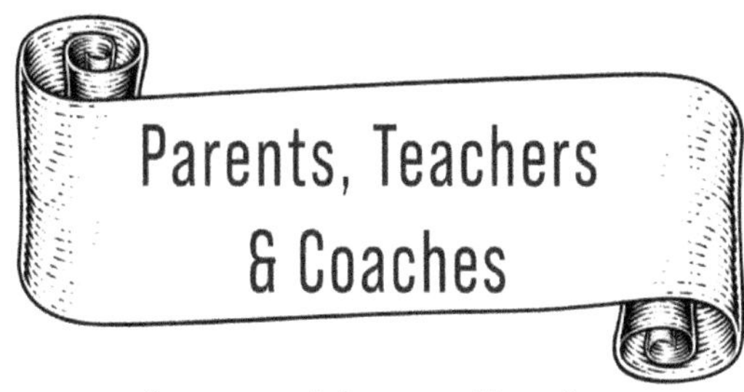

Core message to be conveyed to youth:

Go to the end of your life to determine what truly matters.

Explore innovative ways to have your children, students or athletes repeatedly uploald this message by:

1. Having them read about it
2. Having them write about it.
3. Having them talk about it.
4. Having them think about it.
5. Having them visualizing themselves being it.
6. Having them coach themselves with self-talk.
7. Having them act the message out in behavior.

CONCLUSION

Ubuntu and Humanity's Deep-seated Connection

Ubuntu, stemming from the Nguni Bantu languages of Southern Africa, is not merely a word; it encapsulates an entire worldview and philosophy that defines the African perception of community and connection. Rooted deep in African traditions, Ubuntu stands as a testament to the realization that our individual well-being is inextricably intertwined with the well-being of others. At its core, Ubuntu stresses the importance of universal human bonds, advocating that we're all interconnected in the grand tapestry of life.

To truly understand the profound essence of Ubuntu, one must delve deeper than its direct translation. While it may be rendered as "I am because we are," this phrase barely scratches the surface of its multifaceted meaning. It's a concept that transcends language barriers, urging us to look beyond ourselves and acknowledging the shared experiences, emotions, and destinies that bind humanity together.

This idea of collective belonging is especially significant in today's era of individualism and self-centered pursuits. In a world increasingly marked by division, competition, and isolation, Ubuntu serves as a timely reminder of the power and necessity of human connectivity. It beckons us to remember that no one exists in isolation. Our successes, failures, joys, and sorrows are all intertwined, often rippling out and impacting those around us in ways we might never fully comprehend.

"Everyday Ubuntu," penned by Mungi Ngomane, offers a nuanced exploration of this philosophy. In her insightful narrative, Ngomane eloquently conveys how Ubuntu manifests in daily life, reinforcing that it's not just a theoretical construct but a practical guide to living. Through her words, readers are invited to witness how treating others with dignity, respect, and compassion isn't just an ethical duty but the bedrock of harmonious societies.

One might question why such a concept, which seems innate to human nature, needs to be emphasized. While many of us inherently understand the value of unity and community, modern life often steers us away from such ideals. The frenetic pace of life, the constant chase for materialistic gains, and the pressures to stand out often eclipse the simple truths that Ubuntu highlights. This is where Ngomane's book becomes vital, serving as both a reminder and a manual to realign our compass towards a more compassionate and connected existence.

But how does one live by Ubuntu in everyday life? Ngomane's writings suggest it starts with genuine empathy. It's about feeling the pain of another as if it were your own, rejoicing in their happiness as if it were your triumph. This empathy extends beyond human connections. It's an understanding that every action we take, whether towards other humans, animals, or the environment, has consequences that reverberate through the universe.

Another cornerstone of Ubuntu is humility. Recognizing that we are part of a larger whole and that our individual achievements are often the result of collective effort is vital. This acknowledgment enables us to approach others with an open heart and mind, devoid of judgment or preconceptions.

Furthermore, Ubuntu encourages active listening. In a world where everyone is eager to voice their opinions, taking a step back to truly listen becomes a revolutionary act. By immersing ourselves in the stories and experiences of others, we not only broaden our horizons but foster genuine bonds that stand the test of time.

In the realm of reconciliation, Ubuntu shines especially bright. It promotes forgiveness, understanding that holding onto grudges and animosities harms us more than it harms the perpetrator. While it doesn't negate the importance of justice, it emphasizes the healing power of compassion and unity.

Ultimately, Ubuntu is a celebration of humanity in its purest form. It's an acknowledgment that while our paths may be diverse, our hearts beat with the same rhythm. As Mungi Ngomane beautifully articulates in "Everyday Ubuntu," embracing this philosophy isn't about negating individuality but realizing that our true potential, both as individuals and as a collective, is unlocked when we recognize, appreciate, and nurture the threads of commonality that bind us.

As the world continues to evolve, the teachings of Ubuntu remain timeless, reminding us of the age-old wisdom that in unity lies strength, and in compassion lies the essence of being truly human.

> *"I am because you are........*
>
> *Ubuntu exists when people unite for a common cause..... If we join together, we can overcome our differences and our problems. Whoever we are, wherever we live, whatever our culture, Ubuntu can help us coexist in harmony and peace."*
>
> *"If you want to go quickly, go alone. If you want to go far, go together."*
>
> - African Proverb

"It is only in the heart that one can see rightfully. What is essential is invisible to the eye."

- Antoine de Saint Exupery

"We are not human beings having a spiritual experience; we are spiritual beings having a human experience."

- Pierre Teilhard De Chardin

"Won't you Be My Neighbor?

I've always wanted a neighbor like you.

I've always wanted to

live in a neighborhood

with you. Would you be

mine?

Could you be mine?

Won't you be my neighbor?"

- Fred Rogers

From all about "Me" to all about "You"

Contemplate these:

- Who has vested themselves in your goodness?
- Who has taught you to look for goodness in others, not their failings?
- Who has helped you understand that your treatment of others defines your goodness?
- Who has helped you understand that the best part of you is your goodness toward others?

Sample Parental Training Inputs for Modifying the Flawed Instinct Through Talking (Language)

1. **Early Childhood sharing messages**

   "Sharing with your brother and sister makes all of us proud of you as a person."

   "You make me so happy when you share."

   "Your goodness as a person shines through when you share." "When you share it shows you care about others."

   "I love it when you take joy in sharing with others."

2. **Childhood Temper Tantrums**

   "I know you are upset but we are in a public restaurant and you are making those around us very uncomfortable. You simply cannot disturb others like this. If you don't stop, I will have to take you outside and we will both miss dinner. I'm very hungry and I'm sure you are too but it's disrespectful to act this way in a public place."

   Child continues to throw a tantrum

   "Ok! Out you and I must go. Now that we are outside and can no longer disturb those who are eating, let's take a few minutes to settle down. Try taking some deep slow breaths. Great! Can you tell me why you are so upset? If you settle down, we'll go back and have a nice dinner with your Dad and sister. If not, you and I will sit in the car until your Dad and sister finish their dinners. How would you like it if someone disturbed your dinner by throwing a temper tantrum?"

3. **Sibling Rivalry**

"Remember, you are in this world to help your brother become a better person and he is in this world to help you become a better person. Good brothers help each other."

"Being kind to your sister is priority #1 in this family."

"Being a good person in this family means helping your brother and sister, not hurting them."

4. **Twice a year you have to give away 3 toys to others.**

"It's now time again to share what you have with others. You get to choose what you hope will bring joy to someone else."

"Everything you have is a gift. Sharing your gifts brings happiness to you and to those who receive them."

"What's fair is to share. Sharing ignites the goodness in others."

5. **Tell me how your brother/sister is feeling.**

"What is your sister feeling right now? Have you ever felt like that? How would you like to be treated when you feel like that?"

"What does it feel like when someone hurts your feelings? How can you help others when it happens to them?'

"Have you ever had a good friend hurt your feelings? Why do people hurt the feelings of others?"

6. **One act of kindness per week-required.**

"What are some ways to show kindness at home?" "What are some ways to show kindness at school?"

"People who are kind to others are happier. Why do you think that is the case?"

"Explain, the more you do for others the more you do for yourself."

7. **How would you like to be treated in this situation?**

"When you make a stupid mistake."
"When you do poorly on a test."
"When you have lost your confidence."
"When you are tired and sad."
"When you perform poorly in your sport."

8. **What have you done for others today?**

"What are the little things?"
"What are the hard things?"
"What did you do because you had to?"
"What did you do on your own?"

9. **Who are the kindest people you know?**

"List each of them and explain why you chose each one."

10. Where is the goodness in a person you dislike?

"List 3 good things about the person you dislike."
"There is goodness in everyone but we must look for it intentionally. Give an example."
"Describe the good things you are most proud of in yourself."
"Can you do good things for someone you dislike?"

11. Think of someone who has been rejected by friends or has been bullied.

"How would you feel if you were in their shoes?"
"Has this ever happened to you? If so, how did you feel?"

12. There are givers and takers in life.

"Do you believe givers or takers will find more happiness in life? Why?"
"Are you more of a giver or taker in life?"

13. Individual sports participation.

"How can you do good for others in an individual sport?"
"How can you do good for others in a team sport?"

"What is the value of competitive sport in the context of doing good for others?"

14. Doing good.

"How do you feel when you do good for others?"

"How do they feel?"

"What's the highest good you can do on this planet?"

15. Differences.

"If we are all part of the body of mankind, how should we look at differences?"

"What are the similarities in all of us?"

Conclusion:

Whatever goodness you have within you is there because of others. All goodness is learned. Our past actions and behaviors brought us to our current position, but the same approaches won't be effective for future progress. The best way forward for humanity is to prioritize cooperation, empathy, and kindness over aggression.

"Life is like a tennis game, you can't win without serving." Arthur Ashe

"Coming together is a beginning, staying together is progress, and working together is success." Henry Ford

POSTSCRIPT
(Relevant Questions and Answers)

Question 1.

How can we foster love and compassion for others when we must kill another living life form every day to survive?

None of us had any role in determining how life was to be perpetuated on planet Earth. From a single cell protozoa to Homo sapiens, life is sustained only by consuming other life forms. Every day, we must decide, consciously or not, that my life is more important than the life I'm about to take to sustain mine.

All animals, including Homo sapiens, can be considered living graveyards of all the creatures consumed to support their continued existence. We become numb to killing living creatures, e.g., cattle, fish, birds, deer, elk, buffalo, etc. It's clear that Homo sapiens has been extraordinarily successful in killing other life forms to perpetuate its species.

The more compassion and empathy one has for the animal about to be slaughtered, the more conflicted one will be about taking its life to preserve your own. Slaughterhouses, harvesting factories, and the like remove the killing from the public eye, which helps to minimize the conflict for most people.

Compassion and empathy are powerful deterrents to hurting or killing others, both human and non-human, and both are learned and strengthened in the same physical muscles of the body: Repeated Energy Investment.

In his book "The Science of Evil," Simon Baron-Cohen, Professor of Developmental Psychopathology and Psychiatry at the

University of Cambridge, contends that empathy itself is the most valuable resource for combating evil in our world.

Question 2.

What is **collective efficacy,** and can that be a partial answer to reducing aggression and violence in communities?

Collective efficacy is a term used primarily by sociologists. When communities have a strong sense of trust, solidarity, and commitment to a neighborhood cause, such as reducing aggression and violence, a significant reduction in those behaviors has been witnessed. When community members share a common neighborhood concern for reducing certain behaviors, increased behavioral control in those areas can be the outcome.

Question 3.

If someone is to do harm to me or my family, must I submit if my highest priorities are kindness, love, and compassion?

There is nothing disingenuous about defending yourself or your family and still being a loving, compassionate person. You will hopefully use the least required force to resist the threat, but you have the right to contain the attack by whatever means necessary.

Question 4.

What precautions must I take to prevent the inherited survival mandate from dominating all the moral decisions I make?

Remember, your superbrain is to serve you; you do not serve it! Just like the software on a computer requires periodic upgrades to function properly, so does your brain. Send regular messages to your brain that serve to replace the survival mandate with the mandate to show kindness, empathy, and caring for others, above all else.

Question 5.

What can be done to prevent my reality-distorting, fiction-making brain from sabotaging my moral judgment?

Make sure your brain gets the message that ***truth*** *is more important to you than protecting you from pain. Send the message loud and clear, over and over:* ***Don't hijack the truth to make me feel better in the moment or preserve my moral self-image.***

Ultimately, your treatment of others should determine your goodness as a person. Period!

Question 6.

I have very strong political beliefs and am 99% certain that my views are what's best for our country. I am willing to go to war to fight for what I believe is right. Am I wrong to think this way?

Politics and religion are equally volatile topics. Both have resulted in horrific wars and violent conflicts, all in the name of right. You may be justified in going to war, but the justification should never be to defend your political or religious beliefs themselves. The only justification that would make any moral sense would be to protect the lives and safety of others being persecuted by a particular political regime or religious sect, such as Hitler's death camps. Protecting life, not taking it, should be a central tenet of any political or religious system. Kindness, caring, empathy, fairness, and freedom should be the highest priorities.

Tragically, ideologies are beliefs that often foster oppression.

Question 7.

What's the distinction between *Homo sapiens* and *The human race*?

Homo sapiens refers to a specific species. Human is not a scientific word. Race typically refers to people who are descendants of a common ancestry. Race can be used to divide people on the basis of physical characteristics. Genetic variations between individuals can be as great as those between different races. Three things are important: (1) We are all descendants of the same species; (2) Any reference to race that divides us fuels the moral flaw that has caused unimaginable pain and suffering; and (3) Differences should be celebrated and appreciated to suppress sapiens' moral flaw.
In the words of David Livingstone Smith, author of On Inhumanity, "Dividing human beings into races -our kind and their kind- is the first step on the road to dehumanizing them."

Question 8.

How can we resist dehumanizing those who are different or those who pose a threat of some kind?

Everyone possesses the predisposition to dehumanize others when it serves their purpose. History conclusively demonstrates that we are all vulnerable. If we do not maintain constant vigilance, the tragic predisposition will surface and can be used to sanction almost any inhumane action to contain the perceived threat.

Question 9.

If I agree with the urgent call to action articulated in this book, what practical steps can I take to help with the movement?

Begin immediately presenting the message to friends, neighbors, school boards, churches, and any organizations you have access to.

Here are a few of the organizations in America:

1. *YMCA of America*
2. *National Parent Teacher Association*
3. *National Education Association*
4. *National Association for the Education of Young People*
5. *Big Brothers/Big Sisters of America*
6. *Head Start*

Here are a few International organizations:

1. *Association for Childhood Education International*
2. *International Youth Health Organization*
3. *The Global Coalition for Youth Mental Well-being*
4. *Youth For Change*
5. *World Health Organization Youth Engagement*
6. *The International Association of Adolescent Health*

Question 10.

What is the chance that sapiens' moral flaw can be suppressed in time to save the species from destroying itself?

Without a concerted worldwide effort, the odds are not favorable. My hope is that this book will help bring attention to the flaw and be instrumental in launching a worldwide movement to address it.

Question 11.

What would you recommend as the name for the worldwide movement?

*Because war, slavery, prejudice, genocide, gangs, religious conflicts, etc., and nearly all forms of dehumanization involve some form of aggression toward others, I believe the name **The World Against Aggression (WAA)** works best.*

Question 12.

If some countries of the world begin initiating national programs to curb aggression in their citizens, particularly helping children and young adults become less violently aggressive, won't that make those countries more vulnerable to acts of predatory aggression from countries that don't buy in?

Clearly, that is a risk, but unless sapiens' aggressive nature is replaced with kindness, empathy, caring, compassion, and love for all of its species, the hope for mankind is bleak at best. The greatest risk is to do nothing, leaving the door wide open to nuclear war.

The World Against Aggression (WAA)

Selected References and Readings

PART I

1. Harari, Yuval Noah. *Sapiens: A brief history of humankind.* Random House, 2014.

2. McHenry, Henry. "Skhūl." *Encyclopedia Britannica*, 17 Nov. 2014, https://www.britannica.com/place/Skhul. Accessed 14 February 2023.

3. Bate, Dorothea Minola Alice, and Dorothy Anne Elizabeth Garrod. *The Stone Age of Mount Carmel: Report of the Joint Expedition of the British School of Archaeology in Jerusalem and the American School of Prehistoric Research, 1929-1934. Excavations at the Wady El-Mughara.* AMS Press, 1937.

4. Hoffecker, John F. "The spread of modern humans in Europe." *Proceedings of the National Academy of Sciences* 106.38 (2009): 16040-16045.

5. Fields, R. Douglas. "The roots of human aggression: experiments in humans and animals have started to identify how violent behaviors begin in the brain." *Scientific American* 320.5 (2019): 65.

6. Hamburg, D. "Interview (1998) Conversations with History." Institute of International Studies, U.C. Berkeley, March 2, 1998.

7. Wrangham, Richard W. "Two types of aggression in human evolution." *Proceedings of the National Academy of Sciences* 115.2 (2018): 245-253.

8. Cashdan, Elizabeth, and Stephen M. Downes. "Evolutionary perspectives on human aggression: Introduction to the special issue." *Human Nature* 23 (2012): 1-4.

9. Waterman, H. "Are Humans Inherently Violent? What an Ancient Battle Site Tells Us." *Discover, May* 17 (2019).

10. LeBlanc, Steven A., and E. Katherine. "Register, Constant Battles: The Myth of the Peaceful, Noble Savage." (2003): 78.

11. Davie, Maurice R. *The evolution of war: A study of its role in early societies*. Courier Corporation, 2003.

12. Keeley, Lawrence H. *War before civilization*. OUP USA, 1996.

13. Guilaine, Jean, and Jean Zammit. *The origins of war: Violence in prehistory*. John Wiley & Sons, 2004.

14. Wrangham, Richard. *The goodness paradox: The strange relationship between virtue and violence in human evolution*. Vintage, 2019.

15. Sapolsky, Robert M. *Behave: The biology of humans at our best and worst*. Penguin, 2017.

PART II

1. Riek, Blake M., Eric W. Mania, and Samuel L. Gaertner. "Intergroup threat and outgroup attitudes: A meta-analytic review." *Personality and social psychology review* 10.4 (2006): 336-353.

2. O'Connell, Robert L. *Of arms and men: A history of war, weapons, and aggression*. Oxford University Press, 1990.

3. Brewer, Marilynn B. "The psychology of prejudice: Ingroup love and outgroup hate?." *Journal of social issues* 55.3 (1999): 429-444.

4. Grossman, Dave. *On killing: The psychological cost of learning to kill in war and society*. Open Road Media, 2014.

5. Gay, William C. "The language of war and peace." *Encyclopedia of violence, peace, and conflict* 2 (1999): 303-312.

6. Bourke, Joanna. *An intimate history of killing: Face-to-face killing in twentieth-century warfare*. Basic Books, 1999.

7. Ehrenreich, Barbara. *Blood rites: Origins and history of the passions of war*. Granta Books, 2011.

8. Turiel, Elliot. "Commentary: The problems of prejudice, discrimination, and exclusion. "*International journal of behavioral development* 31.5 (2007): 419-422.

9. Hacker, J. David. "From '20. and odd' to 10 million: the growth of the slave population in the United States." *Slavery & abolition* 41.4 (2020): 840-855.

10. Gates, H.L. "Slavery, by the Numbers." *The Root*: Feb.10, 2014.

11. Brasfield, Curtis G. "Tracing Slave Ancestors: Batchelor, Bradley, Branch, and Wright of Desha County Arkansas." *National Genealogical Society Quarterly* 92 (March 2004): 6-30.

12. University Press of Virginia, Charlottesville, VA. *Afro-American Sources in Virginia: A Guide to Manuscripts* (http://www.upress.virginia.edu/Plunkett/mfp.html)., Michael Plunkett, Editor, and Guide to African American Documentary Resources in North Carolina, Timothy D. Pyatt, Editor.

13. Woodtor, Dee Parmer. *Finding a Place Called Down home: A Guide to African American Genealogy and Historical Identity*. New York: Random House, 1999.

14. Ruffin, C. Bernard, Ill. "In Search of the Unappreciated Past: The Ruffin-Cornick Family Virginia." *National Genealogical Society Quarterly* 81 (June 1993): 126-138.

15. https:/www.quora.com/how-many-people-have-died-in-all-of-the-human-wars

16. https://en.wikipedia.org/wiki:List-of-wars-death-toll

17. https://www.statista.com/statistics/1009819/total-us-military-fatalities-in-american-wars-1775-pres ent/

18. Fischer, Hannah. "US Military casualty statistics: operation new dawn, operation Iraqi freedom, and operation enduring freedom." Library of Congress Washington DC Congressional Research service, 2013.

19. Fischer, Hannah. "A Guide to US Military Casualty Statistics: Operation Freedom's Sentinel, Operation Inherent Resolve, Operation New Dawn, Operation Iraqi Freedom, and Operation Enduring Freedom." LIBRARY OF CONGRESS WASHINGTON DC, 2015.

20. Meierhenrich, Jens. "How many victims were there in the Rwandan genocide? A statistical debate." *Journal of Genocide Research* 22.1 (2020): 72-82.

21. "Commemoration of International Day of Reflection on the 1994 Genocide against the Tutsi in Rwanda - Message of the UNOV/ UNODC Director-General/ Executive Director" United Nations: Office on Drugs and Crime. Retrieved 18 January 2021.

22. Smith, David Livingstone. *Less than human: Why we demean, enslave, and exterminate others*. St. Martin's Press, 2011.

23. Guichaoua, André. "Counting the Rwandan victims of war and genocide: Concluding reflections." *Journal of genocide research* 22.1 (2020): 125-141.

24. Hatzfeld, Jean. *Machete season: The killers in Rwanda speak*. Macmillan, 2005.

25. Waller, James E. *Becoming evil: How ordinary people commit genocide and mass killing*. Oxford University Press, 2007.

26. Hamburg, David A. *Preventing genocide: Practical steps toward early detection and effective action*. Routledge, 2015.

27. Koonz, Claudia. *The nazi conscience.* (2003): 225-246.

28. Beevor, Antony. "They raped every German female from eight to 80." *The Guardian* 1.5 (2002): 02.

29. Roth, John K. "United States Holocaust Memorial Museum." *Holocaust Encyclopedia* (2005).

30. Sharma, Vivek Swaroop. “What Makes a Conflict Religious?” *The National Interest* 154 March/April (2018)

31. Andrew Holt Ph. D (8 November 2018). Religion and the 100 Worst Atrocities in History (apholt.com)

32. Cassidy-Welch, Megan. "Sohail H. Hashmi, ed.: Just Wars, Holy Wars, and Jihads: Christian, Jewish, and Muslim Encounters and Exchanges. Oxford: Oxford University Press, 2012; pp. xvi+ 416." (2014): 602-603.

33. Johnson, James Turner. *Holy war idea in Western and Islamic traditions*. Penn State Press, 2010.

34. Kirsch, Jonathan. *God against the Gods: The history of the war between monotheism and polytheism*. Penguin, 2005.

35. National Drug Intelligence Center, Attorney General's Report to Congress on the Growth of Violent Street Gangs in Suburban Areas. April 2008 (https://www.justice.gov/archive/ndic/pubs27/27612/estimate.htm)

36. Peterson, Dana, Terrance J. Taylor, and Finn-Aage Esbensen. "Gang membership and violent victimization." *Justice Quarterly* 21.4 (2004): 793-815.

37. Bjerregaard, Beth. "Self-definitions of gang membership and involvement in delinquent activities." *Youth & Society* 34.1 (2002): 31-54.

38. Miller, Walter Benson. *The Growth of Youth Gang Problems in the United States, 1970-98: Report*. US Department of Justice, Office of Justice Programs, Office of Juvenile Justice and Delinquency Prevention, 2001.

39. Thrasher, Frederic Milton. *The gang: A study of 1,313 gangs in Chicago*. University of Chicago press, 2013.

PART III

1. Campbell, Joseph. *The hero with a thousand faces*. Vol. 17. New World Library, 2008.

2. Campbell, Joseph. *The hero's journey: Joseph Campbell on his life and work*. Vol. 7. New World Library, 2003.

3. Campbell, Joseph, and Bill Moyers. *The power of myth*. Anchor, 2011.

4. Huang, Niwen, et al. "The dark side of malleability: Incremental theory promotes immoral behaviors." *Frontiers in psychology* 8 (2017): 1341.

5. Buchtel, Emma E., et al. "Immorality east and west: Are immoral behaviors especially harmful, or especially uncivilized?." *Personality and Social Psychology Bulletin* 41.10 (2015): 1382-1394.

6. Ariely, Dan, and Simon Jones. *The honest truth about dishonesty*. New York: HarperCollins Publishers, 2012.

7. Loehr, James. "Forces and Factors that May Corrupt Your Moral Reasoning and Judgement." in *Leading with Character*: Wiley Publishing. P. 171-173, 2021.

8. Mele, Alfred R., and H. William. *Self-deception unmasked*. Vol. 6. Princeton University Press, 2001.

9. M. Cazzaniga, *The Ethical Brain.* Chicago: University of Chicago Press, 2005.

10. Trivers, Robert. *Natural selection and social theory: Selected papers of Robert Trivers.* Oxford University Press, 2002.

11. Baard, Erik. "The guilt-free soldier." *Village Voice* 28 (2003).

12. Mackie, John. *Ethics: Inventing right and wrong.* Penguin UK, 1990.

13. Gino, Francesca. "Understanding ordinary unethical behavior: Why people who value morality act immorally." *Current opinion in behavioral sciences* 3 (2015): 107-111.

14. Tenbrunsel, Ann E., and David M. Messick. "Ethical fading: The role of self-deception in unethical behavior." *Social justice research* 17 (2004): 223-236.

15. Tenbrunsel, A., and Messick, D. "We Are Creative Narrators," Ethical Fading: The Role of Self-Deception in Unethical Behavior'. *Social Justice Research* 1 7: 2 2 5, 2014.

16. Mele, Alfred R., and H. William. *Self-deception unmasked.* Vol. 6. Princeton University Press, 2001.

17. Gazzaniga, Michael S. *The ethical brain.* Dana press, 2005.

18. Trivers, Robert. "Deceit and self-deception." *Mind the gap: Tracing the origins of human universals* (2010): 373-393.

19. Mackie, John. *Ethics: Inventing right and wrong.* Penguin UK, 1990.

20. Slovic, Paul. Moral Numbing. "Psychic numbing and genocide." *Psychological Science Agenda.* American Psychological Association (November) 2007.

PART IV

1. Ahmad, Hafiz Ishfaq, et al. "The domestication makeup: Evolution, survival, and challenges." *Frontiers in Ecology and Evolution* 8 (2020): 103.

2. Wrangham, Richard W. "Hypotheses for the evolution of reduced reactive aggression in the context of human self-domestication." *Frontiers in Psychology* (2019): 1914.

3. Theofanopoulou, Constantina, et al. "Comparative genomic evidence for self-domestication in Homo sapiens." *BioRxiv* (2017): 125799.

4. Leach, HelenM. "Human domestication reconsidered." *Current anthropology* 44.3 (2003): 349-368.

5. MacLean, Evan L., et al. "The evolution of self-control." *Proceedings of the National Academy of Sciences* 111.20 (2014): E2140-E2148.

6. Henrich, Joseph. "The secret of our success." *The Secret of Our Success: How culture is driving human evolution, domesticating our species, and making us smarter*. Princeton University Press, 2016.

7. Hare, Brian. "Survival of the friendliest: Homo sapiens evolved via selection for prosociality." *Annual review of psychology* 68 (2017): 155-186.

8. Hartman, Steve. Kindness 101 for Teachers, CBS On the Road, CBS Evening News

9. Benítez-Burraco, Antonio, Zanna Clay, and Vera Kempe. "Self-domestication and human evolution." *Frontiers in Psychology* 11 (2020): 2007.

10. Zeder, Melinda A. "Domestication: Definition and overview." *Encyclopedia of global archaeology*. Cham: Springer International Publishing, 2020. 3348-3358.

11. Zeder, Melinda A. "The domestication of animals." *Journal of anthropological research* 68.2 (2012): 161-190.

12. Price, Edward O. "Behavioral aspects of animal domestication." *The quarterly review of biology"* 59.1 (1984): 1-32.

13. Telechea, F. (2015). "Domestication and genetics." In Pontarotti, P. (ed.). *Evolutionary Biology: Biodiversification from Genotype to Phenotype*. Springer. p. 397.

14. Darwin, Charles. *The variation of animals and plants under domestication*. Vol. 2. J. Murray, 1868.

15. Larson, Greger, and Dorian Q. Fuller. "The evolution of animal domestication." *Annual Review of Ecology, Evolution, and Systematics* 45 (2014): 115-136.

16. Moore, David S. *The developing genome: An introduction to behavioral epigenetics*. Oxford University Press, 2015.

17. Berger, Shelley L., et al. "An operational definition of epigenetics." *Genes & development* 23.7 (2009): 781-783.

18. Nestler, Eric J. "Transgenerational epigenetic contributions to stress responses: fact or fiction?." *PLoS biology* 14.3 (2016): e1002426.

19. "Epigenetics" (http://www.bio-medicine.org). Bio-Medicine.org. 21 May 2011.

20. "Epigenetic Mechanisms - an overview | ScienceDirect Topics" (https://www.sciencedirect.com/topics/biochemestry-genetics-and-molecular-biology/epigenetic-m echanism). www.sciencedirect.com. February, 2012.

21. Bošković, Ana, and Oliver J. Rando. "Transgenerational epigenetic inheritance." *Annual review of genetics* 52 (2018): 21-41.

22. Meisenberg, Gerhard. “Repressor Protein.” ScienceDirect, *Physiological Genetics*, 1979.

23. Brackett, Marc. *Permission to feel: Unlocking the power of emotions to help our kids, ourselves, and our society thrive*. Celadon Books, 2019.

24. Ngomane, Nompumelelo Mungi. *Everyday ubuntu: Living better together, the African way*. Random House, 2019.

25. De Chardin, Pierre Teilhard. *The phenomenon of man*. Lulu Press, Inc, 2018.

26. “30 Organizations Working to End Hunger.” *HUMAN RIGHTS CAREERS*, https://www.humanrightscareers.com/magazine/organizations-end-hunger

27. “Losing 25,000 to Hunger Every Day.” *United Nations UN Chronicle*, https://www.un.org/en/chronicle/article /losing-25000-hunger-every-day

28. “9 million people die every year from hunger, WFP Chief tells Food System Summit.” *World Food Program*, https://www.wfp.org/news/world-wealth-9-million-people-die-every-year-hunger-wfp- chief- tells-food-system-summit

29. “How Many People Die From Hunger Each Year.” *The World Counts*, https://www.theworldcounts.com/challenges/people-and-poverty/hunger-and-obesity/how-many- people-die-from-hunger-each-year

POSTSCRIPT

1. Baron-Cohen, Simon. *The science of evil: On empathy and the origins of cruelty*. Basic books, 2012.

2. Sampson, Robert J., Stephen W. Raudenbush, and Felton Earls. "Neighborhoods and violent crime: A multilevel study of collective efficacy." *science* 277.5328 (1997): 918-924.

3. Smith, David Livingstone. *Less than human: Why we demean, enslave, and exterminate others*. St. Martin's Press, 2011. p 269.

4. Smith, David Livingstone. *On inhumanity: Dehumanization and how to resist it*. Oxford University Press, 2020. p 41.

Acknowledgments

First to my son Patrick Loehr for his expertise in producing this manuscript. Without his knowledge, technical skills, and encouragement, this book could not have happened.

To all who provided feedback on the manuscript: Mike Loehr, Jeff Loehr, Gordon & Rehana Uheling, John Collingwood, Sheila Ohlsson, Caren Kenney, Virginia Savage, Chris Osorio, Michael Rouse, Kelsey Abergel, Paul Hancock, and Lee DeYoung.

Thank you!

About the Author

Jim Loehr holds both a doctorate and a master's degree in psychology. He has been inducted into four halls of fame and has written 19 books. Among them is his latest, "Wise Decisions: A Science-Based Approach to Making Better Choices," co-authored with Dr. Sheila Ohlsson. Additionally, Jim co-wrote the national bestseller, "The Power of Full Engagement." He is a pioneer in applying psychology to improve human performance.

Visit Jim-Loehr.com for more information.